Edmund W. Schuster
Stuart J. Allen
David L. Brock

Global RFID

The Value of the EPCglobal Network
for Supply Chain Management

With 33 Figures and 8 Tables

 Springer

Edmund W. Schuster
Dr. David L. Brock
MIT
77 Massachusetts Ave., Rm 35-212
Cambridge, MA 02139
USA
edmund_w@mit.edu
dlb@mit.edu

Dr. Stuart J. Allen
7904 60th Drive NE
Marysville, WA 98270
USA
stuart99allen@yahoo.com

Library of Congress Control Number: 2006936845

ISBN 978-3-540-35654-7 Springer Berlin Heidelberg New York

Springer is part of Springer Science+Business Media

springer.com

Production: LE-TEX Jelonek, Schmidt & Vöckler GbR, Leipzig
Cover-design: WMX Design GmbH, Heidelberg

SPIN 11784760   134/3100YL - 5 4 3 2 1 0   Printed on acid-free paper

Table of Contents

List of Figures and Tables

FOREWORD

Who Invented the EPC?

by Kevin Ashton

One of the questions I get asked most often is "Who invented the Electronic Product Code?" I know I am expected to answer with a single name – some heroic figure who labored completely alone, perhaps me, perhaps someone else. But that's never how invention works, and it's certainly not true of the EPC™. Many hundreds if not thousands of people deserve credit.

The EPC system is just a way of using RFID – radio-frequency identification. The name RFID dates back to the early 1990s. The technology has roots in radar-related work done in the 1940s. During these six decades, hundreds of engineers and scientists, most of them unknown, worked on RFID. By the time I first saw it in the mid-1990s it was commercially available, used for applications like controlling access to office buildings and automating toll collection. The same is true of Sanjay Sarma, David Brock and Sunny Siu, the three people with whom I cofounded MIT's Auto-ID Center. If any of us "discovered" RFID, it was only by looking in our pockets or at the windshields of our cars.

The big challenge in the mid-1990s was to make RFID cheap and standard enough that it could be used everywhere, and to find the killer applications that would make ubiquity useful – to build a mass market for what, until then, was a niche technology.

David Brock, a research scientist in MIT's Artificial Intelligence Lab, wanted to use RFID to help robots interact with the world around them. He saw that combining the automatic identification capability of RFID with the limitless information access of the Internet could be a powerful, practical way to do this. Sanjay Sarma, a rising Professor in MIT's Mechanical Engineering department, championed Brock's idea and helped develop it. Sarma also reached out to his faculty colleague Sunny Siu for additional expertise on the networking challenges.

At the same time, I was a junior Brand Manager at Procter & Gamble worried about a much more mundane problem: how to keep my products on the shelf. Embedding RFID tags in the products, and RFID readers in the shelf, seemed like the perfect – indeed the only – way to do this. But I needed RFID to be cheaper, better, and standardized in an open system. In early 1999, by sheer chance, I met Brock and Sarma. The result was a potent meeting of minds. I was looking to fund research, and Brock, Sarma and Siu were looking for research funding.

Working with Alan Haberman of the Uniform Code Council, one of the founding fathers of the UPC bar code, and Allan Boath of the Gillette Company, we developed a plan for a new industry funded research consortium at MIT. Haberman wanted to call it the Center For Automatic Identification And Data Capture. At the last minute I persuaded him to abbreviate it to the Auto-ID Center. But my luck with names is hit and miss: inspired by the bar code, I had the bad idea of calling Auto-ID Center's technology UPC2. Brock and Sarma saved the day – one of them, I cannot remember which, proposed a far better alternative: EPC, for electronic product code.

The Auto-ID Center opened on October 1, 1999. P&G loaned me to MIT to act as Executive Director, and Sunny Siu was the first Research Director. When Sunny left MIT in 2000, Sanjay Sarma, always the guiding light behind the research, took on his role.

The four-year project was more challenging and ultimately more fruitful than any of us expected. The Center grew from its founding three sponsors, P&G, Gillette and the UCC, to 103 companies. It expanded globally, funding additional research labs in Japan, Australia, China, Switzerland and the United Kingdom. Working closely with industry, it published hundreds of research papers by hundreds of researchers, and solved many of the problems standing in the way of low cost, high performance, ubiquitous RFID.

In 2003 the Center's sponsors were ready to use the EPC. MIT licensed the technology to the UCC, which established a new subsidiary, called EPC Global, to operate the EPC system all over the world. The labs were renamed Auto-ID Labs and funded by EPC Global to continue advanced research related to the EPC System.

Who invented the EPC? It's the wrong question. Invention is not an activity of individuals. It is the output of many people, spread around the world and across decades, working hard to improve things, sometimes alone, but more often in teams and frequently unaware of one another. The heroic individual inventor is a myth created after the fact – an unfortunate side effect of success. When something succeeds, as the EPC indisputably has, people want to know 'who invented it?' as if they urgently needed to carve a statue. But no *one* invented the EPC system. It was and still is the art of many – a brilliant, vibrant society of disparate minds and voices, all working passionately to make something new and important. That may be a more complicated answer, but it is also more uplifting. One of the greatest lessons of the EPC, and of all other technologies if their stories were truly told, is that invention is not a lonely act: it means bridging oceans, generations, and cultures to build a community of creativity.

Kevin Ashton, cofounder and former executive director, MIT Auto-ID Center

A Large-Scale Effort

by Sanjay Sarma

This book chronicles the technological advancement of RFID (Radio Frequency Identification) technology and the EPCglobal Network™. David Brock, one of the authors of this book, was a key part of the EPC research effort, and it is a pleasure to see him and his fellow authors describe the technology and its applications. The EPC was developed with goals both mundane and magnificent: from tagging cases of milk to the vision of an "Internet of Things."

Looking back, many of the predictions made by the Auto-ID Center, which at the time were considered reckless or visionary depending on the perspective, have come true. Others seem to be on the way to coming true. For example, in May 2006, the first press-release announcing 5 cent EPC tags appeared in the commercial media. It is easy to forget how contentious an issue the mere mention of a 5 cent tag was in 2000 and 2001. However, time smoothes even the most intense debate. At the time of writing this Foreword, over 800 companies worldwide are members of EPCglobal, and there have been thousands of EPC site implementations and millions of EPC tags shipped on five continents.

Some of our speculations have also not come true (yet). For example item-level tagging has not taken off as we expected it would. Stay tuned, though, because it is beginning to accelerate in the pharmaceutical, electronics, and media industries almost unnoticed.

The Auto-ID Center was something of a saga in technology development and collaboration. Through all this, careers were wagered and made, companies were floated (though some did not survive), reputations were put on the line, theses were written, and degrees were granted. The people involved need to be recognized, and I attempt to do so below.

Beginning in October 1999, the effort behind the EPC (electronic product code) and RFID rapidly expanded to include 103 sponsors who contributed over $20 million for research and development. As an outgrowth of the original MIT Auto-ID Center, six other academic laboratories quickly formed including (with founders listed) the University of Cambridge, UK (2000, Dr. Duncan McFarlane); the University of Adelaide, Australia (2001, Professor Peter Cole); the University of St. Gallen, Switzerland (2002, Professor Elgar Fleisch); Keio University, Japan (2002, Professor Jun Murai); Fudan University, China (2002, Professor Hao Min); and Information and Communication University, Korea (2005, Professor Sang Gug Lee). These universities are all currently in a confederation called Auto-ID Labs (www.autoidlabs.org), which is funded largely by EPCglobal, Inc. (www.epcglobalinc.org), a non profit, joint venture of GS1 US (formerly known as the Uniform Code Council (UCC)) and GS1 (formerly know as the International Article Numbering Association (EAN)).

Independent of this organization, other universities worldwide are conducting ongoing RFID research along with many large corporations from a variety of industries. Governmental agencies have also done much work.

In October 2003, the standards of the MIT Auto-ID Center were licensed to EPCglobal Inc. for commercial development. Since 2003, companies from several different industries, in conjunction with EPCglobal, have continued to refine the technology for practical application.

The four people who founded the MIT Auto-ID Center were, in addition to me, Dr. David L. Brock and Professor Sunny Siu from MIT, and Kevin Ashton from industry. While it is simpler to attribute the contributions of the academics in idea-generation, technology development, and system engineering, it is sometimes difficult to understand the importance and power of organization and communication. For this, Kevin Ashton, who served as Executive Director, and is now Vice President of Marketing for Thingmagic, deserves a great deal of credit and gratitude from the industry. Through many challenges, the Auto-ID Center was successful largely because of the leadership, teamwork, and interaction between the four of us.

In addition to the founders, several people from industry and academia also contributed a great deal to the formation and expansion of the MIT Auto-ID Center. Alan Haberman of the UCC, who served as the founding Chairman and Dick Cantwell of Gillette, who took over this role in 2001, both added a dimension of credibility, integrity, and leadership that propelled the MIT Auto-ID Center from an academic research project to a force in industry. Their enthusiasm and progressive attitudes toward technology acted as an effective lightning rod to draw attention to the EPC and its potential as an operational tool to create value for business. Noel Eberhardt, who was at a fledgling, but now defunct division of Motorola was a great source of moral support. Finally, Professor Nam Suh, head of the Department of Mechanical Engineering at MIT, was a beacon of support not only by seeding the project financially, but also by encouraging us to take risks.

Several research engineers and students at the MIT Auto-ID Center were instrumental in its early success. They are Joe Foley (who implemented the first version of the EPC System in its entirety), Dan Engels (then still a graduate student in Electrical Engineering), Erik Nygren, Ching Law (whose early paper on anti-collision with Sunny Siu was very influential), Steven Ho (who examined warehouses designed with RFID), Yogesh Joshi (who studied the bull-whip effect with and without RFID-enabled visibility), Bink, Niranjan Kundapur, Yun Kang, Kashif Khan (who studied vibratory assembly of tags), Brendon Lewis, Timothy P. Milne (who was a key player in the development of an integration strategy for EPC), Grabriel Nasser, Prasad Putta, and Sridhar Ramachandran. Also members of this group, the afore-mentioned Dr. Daniel W. Engels and Robin Koh (associate directors), and Tom A. Scharfeld (research manager), continued to conduct research and lead students as part of Auto-ID Labs. Their efforts contributed a great deal to tag protocols (Engels), an understanding of the medical counterfeit problem (Koh), and the development of a certification process along with early RFID research in the US and Japan (Scharfeld).

Christian Floerkemeier of ETH Zurich deserves credit for leading the Product Markup Language (PML) group. Mark Harrison, Steven Hodges, and Allan Thorne of Cambridge were key early contributors in software and hardware thinking. Steve Weis was our security and privacy guru along with his advisor, the famous Professor Ronald Rivest. Elliott Maxwell led the public policy council, which served as the conscience of the MIT Auto-ID Center in matters involving social issues related to the large -scale use of RFID. In particular, Simson Garfinkel deserves special mention as a member of this council and as a prominent writer and journalist in the field.

We had many contributors from the technical community. A number of faculty, friends, and industrial colleagues from the industry were impromptu advisors, including Timo Lindstrom (then head of Rafsec, who with Samuli Stromberg, gave us many insights into tag manufacturing); John Price, Dr. Stephen Smith, and John Rolin of Alien Technology (who gave us a deep understanding of UHF tags and fluidic self-assembly); Andreas Plettner (a founder of Flexchip, a pioneering company that unfortunately did not survive); and Dr. Gitanjali Swamy, our informal expert on VLSI and the economics of the 5 cent tag from the early days until our final paper on the detailed simulation showing that a 5c tag was possible. Dirk Heyman and his colleagues at Sun Microsystems, Inc. were also very influential supporters of the Auto-ID Center in the early days, and Dirk went on to become the chairman of the Technology Board. Silvio Albano led the field trial and was a dynamo in the evangelization of EPC technology.

The birth of EPCglobal from the Auto-ID Center also involved significant work, leadership, and risk-taking on behalf of non-profit groups. From the Uniform Code Council, Bernie Hogan, Mike Di Yeso, and Steve Brown were extraordinary partners. In my view, the saga of their work in forming EPCglobal is an effort about which another book needs to be written. From MIT, Carol Carr and Tom Henneberry were key contributors along with Lita Nelsen.   Aside from their roles in industry, Bruce Delagi of Sun Microsystems and Ken Traub of Connecterra spent a great deal of time driving the standards effort, which eventually blossomed within EPCglobal.

Administrators and program managers involved in the development and execution of the numerous meetings and membership relations efforts of the MIT Auto-ID Center deserve recognition for their long hours of work and dedication. These include Brooke K. Peterson, who served as associate director, Joyce Lo (program manager), Tracy Skeete (events manager), Toni Pommet (program manager – Auto-ID Labs), and Carolyn Skeete (administrative assistant). Finally, there is David Rodriguera who was and remains a key member of the Auto-ID Center and now Labs both as my assistant and as the glue for the students.

As Kevin Ashton mentions above, a number of people from universities and business contributed ideas to the MIT Auto-ID Center through published research. Many of theses papers have become citations included as part of this book, although all of these authors contributed in various ways to the idea of using unique identification as a tool in supply chains. A list of these individuals includes:

Vivek Agarwal, Silvio Albano, Keith Alexander, Dipan Anarkat, Alfred Angerer, KevinAshton, Sana Ayub, Brandon Bean, Attilio Bellman, Anthony Bigornia, Garry Birkhofer, Michael Boushka, Jeffrey D.Brooks, Stephen A.Brown, James Brusey, Brian Cantwell, Carol T.Carr, James Carr, Indy Chackrabarti,Yoon Chang, Gavin Chappell, Chunhong Chen, Xiaojun Chen, Jin-Lung Chirn, Nirav Chokshi, Yvonne Chow, Peter H.Cole, Daniel Corsten, Chris Cummins, Markus Dierkes, Dittmann, Lars, Helen Duce, Robert Dudley, Joe Dunlap, David Durdan, Noel Eberhardt, Daniel W.Engels, Elgar Fleisch, Martyn Fletcher, Rich Fletcher, Christian Floerke-meier, Andrés García, Greg Gilbert, Tig Gilliam, Lyle Ginsburg, Amit Goyal, Kathryn Gramling, Sandra Gross, Chris Grubelic, Jennifer Haber-stroh, Hisakazu Hada, Thaddeus Haffey, David M.Hall, Stephan Haller, Yifeng Han, Mark Harrison, Junius K.Ho, Steve Hodges, Chenling Huang, Behnam Jamali, Dawei Jin, Yogesh V.Joshi, Shang-ling Jui, Ajit Kambil, Yun Y.Kang, Yuusuke Kawakita, Kashif Khan, Mike Kindy, Herb Kleinberger, Robin Koh, Jiro Kokuryo, A. G.Kulkarni, Elaine M.Lai, Nhat-So Lam, Mat-thias Lampe, Ching Law, Kayi Lee, Sang-Gug Lee, Stephen Leng, Kin Seong Leong, Brendon W. Lewis, Qiang Li, Joyce S. Lo, Paschalis Loucaides, Yael Maguire, Hugo Mallinson, Uttara Marti, Andrew McDon-ald, Duncan McFarlane, Florian Michahelles, Timothy P.Milne, Hao Min, Jin Mitsugi, Dhaval Moogimane, Humberto Moran, Jun Murai, Osamu Nakamura, Mun Leng Ng, Mikako Ogawa, Ted Osinski, Ravikanth Pappu, Ajith Kumar Parlikad, Sumukh Pathare, Christian Plenge, Rehmi Post, Kart Prince, Laxmiprasad Putta, Damith C.Ranasinghe, Richard Redemski, Mat-thew Reynolds, Jason Richard, Joseph Richards, Jürgen Ringbeck, Sanjay E. Sarma, Tom A. Scharfeld, Paul Schmidt, Patrick Schmitt, Bernd Schoner, Chris Sheedy, Kai-Yeung Siu, Jeffrey Smith, Simon Smith, Thorsten Staake, Martin Strassner, Stefan Stroh, Gitanjali Swamy, Christian Tellkamp, Frédéric Thiesse, Alan Thorne, Joseph Tobolski, Hideaki Tomikawa, Colin J.Towner, Harry Tsai, Masaki Umejima, Yojiro Uo, Tianyang Wang, Stephen A.Weis, John R.Williams, Jonathan Wolk, Chien Yaw Wong, Maurice Woods, Alia Ahmad Zaharudin, Feng Zhou, Zheng Zhu, Steve Zujkowski.

Our apologies in advance for anyone we have inadvertently not listed. For those interested in reading these early papers, Auto-ID Labs has made them available at www.autoidlabs.org

Special thanks must go to the numerous people who have contributed time, expertise, equipment, encouragement, and funding from the member companies of the MIT Auto-ID Center, including:

Technology Board

Accenture	ACNielsen
Alien	Avery Dennison
AWID	British Telecommunications (BT)
Cash's	Catalina Marketing Corp
Checkpoint Systems, Inc.	ConnecTerra, Inc.
Ember Corporation	Embrace Networks
Flexchip AG	Flint Ink
GEA Consulting	GlobeRanger
IBM Business Consulting Services	IDTechEx
Impinj, Inc.	Information Resources, Inc.
Intel	Intermec
Invensys PLC	Ishida Co, Ltd.
KSW Microtec AG	Manhattan Associates
Markem Corp.	Matrics
Morningside Technologies	NCR Corporation
Nihon Unisys Ltd.	Nippon Telegraph and Telephone Corporation
NTT Comware	OATSystems
Omron Corporation	Philips Semiconductors
Rafsec	RF Saw Components
SAMSYS	SAP
Savi Technology	Sensitech
Sensormatic Electronics Corp	Siemens Dematic Corp.
STMicroelectronics	Sun Microsystems
Symbol Technologies	Toppan Forms
ThingMagic	Vizional Te
Toray International, Inc.	Zebra Technologies Corporation

Ongoing research concerning RFID and the EPCglobal Network continues at MIT through a new entity, Auto-ID Labs (http://autoid.mit.edu/cs/), under the leadership of Professor John R. Williams (Director). This new group includes Dr. Abel Sanchez, Dr. Daniel Engels, Dr. Brian Subirana, Stephen Miles, Sivaram Cheekiralla, Ching-Huei Tsou, and David Rodriguera. Former members of Auto-ID Labs who deserve recognition include Tatsuya Inaba, Robin Koh, and Tom A. Scharfeld.

Through many years of extremely hard work, I am happy to see that the EPCgobal Network and RFID technology are on a firm footing to add value in business. Given my direct experience with the MIT Auto-ID Center, I have found that the complex interaction between theoretical and applied academic research, in combination with industrial application and practical innovation, can be harnessed to create something meaningful and lasting. While it is hard to predict the future, the momentum behind the EPC will most certainly lead to more innovations occurring at an increasingly rapid pace. Being a part of this process has provided special insights to me that will become the focus of some of my future research at MIT.

Professor Sanjay E. Sarma,
cofounder and chairman of research, MIT Auto-ID Labs,
Associate Professor of Mechanical Engineering, MIT

Preface

Edward Gibbon once wrote, "All that is human must retrograde if it does not advance."[1] We believe technological progress is a fundamental aspect of human nature that must be encouraged. Without it, living standards fall and civilizations collapse.

This book examines RFID (Radio Frequency Identification) technology and the EPCglobal Network™ through case studies along with our insights concerning the future influence RFID and the EPC will have on business. An ambitious effort to create value within commerce, the technology focuses on building the capability and infrastructure to achieve unique identification for physical objects on a large scale. The ultimate goal is to create an "Internet of Things" by connecting physical objects to computer networks.

This book introduces the topic of RFID and the EPCglobal Network to a wide audience. In doing so, we focus on the essentials of the technology and a number of application areas. With primary focus on supply chain management (SCM), a significant part of the book analyzes the practical aspects of implementation.

Taking a longer view, the advancement of SCM is very much dependent on data. The wave of the future is the integration of SCM with innovations in computer science and other areas, to create better decision-making and to increase speed. With this perspective, we envision RFID and the EPCglobal Network as basic information technology to support greater integration. To this end, the book details the technological potential along with a balanced appraisal of strengths and weaknesses.

Upon publication of this book, we feel honored to acknowledge a number of people who have contributed in many ways. The nature of our work depends heavily on collaboration with a wide group of colleagues from industry, academia, and governmental agencies, along with student research work. It is only through intellectual discourse that we have been able to continually refine and improve our ideas.

Professional colleagues and friends who have generously shared their ideas with us over the years include Mark Dinning, Pinaki Kar, Professor Masahiro Arakawa (Kansai University, Japan), Tom A. Scharfeld, and Dr. James M. Masters of the RAND Corporation. Parts of this book trace to their contributions in a range of different areas including case studies and overall philosophy of what is relevant in education and research.

We would also like to recognize the cadre of undergraduate and graduate students at MIT (2000 – 2006) who through their insights, enthusiasm, and desire to learn have been an inspiration. The students with whom we have worked as thesis and internship supervisors come from several degree programs at MIT, ranging from Undergraduate (Mechanical Engineering, Aerospace Engineering, Physics) to Graduate (Master of Engineering in Electrical Engineering and Computer Science, EECS; Master of Engineering in Logistics, MLOG; and Master of Science in Systems Design and Management, SDM). These students include Tom Albertson, Attilio Bellman, Chaitra Chandrasekhar, Mary Chang, Indy Chakrabarti, Henry Chen, I-Han Chen, Joseph Dahmen, Mark Dinning, Kevin Emery, Narsi Reddy Gayam, Taehee Han, Pinaki Kar, Elaine M. Lai, Nhat-So Lam, Kuang Yuan Ler, Hongmin Li, Ming Li, Max Locher, Peden P. Nichols, Christian K. (Rocco) Repetski, Adam Schlesinger, Chatchai Unahabhokha, Ping Wang, Kazunari Watanabe, and Marat Zborovskiy.

Preparation of a book is not a singular effort. It takes input from a number of different people. We wish to thank those who have helped with specific comments about content, especially Stefanie Alki Delichatsios, a graduate student at MIT Auto-ID Labs, who provided technical guidance at a critical juncture during the writing of this book. Others also made significant contributions including Timothy J. Kutz, Sr. of MorganFranklin Corporation, Nicholas Fergusson of EPCglobal, Inc., Professor John R. Williams of MIT Auto-ID Labs, and Dee F. Biggs of Welch Foods, a long time colleague, friend, and leader.

From an editorial standpoint, we thank Dr. Niels Peter Thomas of Springer Verlag for his abundant skills in improving the organization and focus of the book, along with help in various administrative and creative matters. Above all, we appreciated Niels' dedication to handling important decisions quickly with insight and decisiveness.

Participation in professional societies always serves as an important part of gaining insight into the big picture surrounding a new technology like RFID. We would like to thank our colleagues in the American Production and Inventory Control Society: the Association for Operations Management (APICS), the Council for Supply Chain Management Professionals (CSCMP), and the Institute for Operations Research and Management Sciences (INFORMS).

In addition, trade publications are an important part of communicating with industry. We would like to thank Mark Roberti, Editor of *RFID Journal,* for his efforts in making information available, Doug Kelly, Editor of *APICS – The Performance Advantage*, for publishing some of our early work concerning RFID to a wide audience in industry, and the Cutter Consortium, a group of over 125 experts led by Karen Coburn (President and CEO), Anne Mullaney (Vice President, Product Development and Marketing), and Professors Robert D. Austin (Harvard Business School) and Richard L. Nolan (University of Washington).

MIT is a rich reservoir of intellectual capital. Academic leadership makes research efforts like the Auto-ID Center and Labs possible. We owe gratitude to Professor Jung-Hoon Chun, who is Director of the MIT Laboratory for Manufacturing and Productivity (LMP); the MIT Center for Transportation and Logistics (CTL) where the first author worked during early stages of RFID applications research; and industrial sponsors MorganFranklin Corporation, LG CNS (Korea), Raytheon, Siemens, and ReadyTouch who are founding members of our new research effort, the MIT Laboratory for Manufacturing and Productivity – Data Center Program (www.mitdata-center.org).

Besides professional relationships, long-time friends and family play an important role in encouragement, support, and honest criticism. We wish to thank Dino and Ruth Lioi and their family, including David, Sue, Dean, Kris, Tia, Dena, Adam, and Peyton; Mark and Diane Riethman and their children Evan and Marissa; Ron and Nancy Rife; Al and Diane Solderitsch, Cassandra Solderitsch, Katrina and John Bradley and their daughter Callah; John Mackin; Pinaki and Dr. Rashmi Kar; Sean Boutin & Family; Mr. Bruce Hamilton; and the best of all friends, Kirby, who possessed "all the virtues of man without his vices."[2]

Finally, for those who make a true difference in our life's we reserve special recognition, especially Edmund Roswell Bennett and his wife Anna for representing the value of education, writing, and the development of applied technology for things that matter; and the memory of Edmund L. and Agnes J. (Sue) Schuster, who were people of strong character, determination, and intellect. They both continue to provide inspiration.

Edmund W. Schuster   *Oktober 2006*
Cambridge, MA, U.S.A

Stuart J. Allen
Erie, PA, U.S.A

David L. Brock
Cambridge, MA, U.S.A

PART I: INTRODUCTION

CHAPTER 1

The Emergence of a New Key Technology

The essence of innovation is the blending of ideas with the science and practice of engineering. Nowhere is this process more active than in the area of identification technologies. Taking full advantage of improvements in microcircuit design and production, computer science, the Internet, and other mechanical technologies, the EPCglobal Network™ seeks to establish a large-scale computer infrastructure for merging data, information, and physical objects together. This will create a new, networked physical world that is similar to the Internet.

The power of the EPCglobal Network involves the combination of knowledge from many different fields ranging from computer science and engineering to supply chain management. Within the next ten years, practitioners and researchers alike envision the EPCglobal Network as becoming the predominant means of object identification within business.

The Bar Code and Beyond

Considered one of the greatest innovations of the 20th century,[1] the bar code represents the first large-scale, automated effort to identify objects. For retailers, the implementation of the bar code has led to improved pricing accuracy, greater labor efficiency, and reduced checkout time for customers.[2] In addition to these operational efficiencies, the bar code has come to represent an icon present on the package of nearly every consumer good marketed in the United States.

Given these early successes, manufacturers began to adopt the bar code as a means of improving inventory accuracy and to coordinate supply chain operations. Product level data on time and place obtained from bar code systems greatly improved temporal and spatial utility within supply chains.[3, 4] During the 1980s, this was an important element in better management of inventory deployed at forward warehouses along with enhanced customer service. The decreasing cost of computing hardware along with improved Enterprise Resource Planning (ERP) systems used to organize and communicate data combined to accelerate the adoption of the bar code by many firms.

Since bar codes are now a mature technology, it is natural to look forward to the next stage in the commercial use of identification systems. The historical focus of bar codes has concentrated on identification of an object type. For a consumer goods item, this means the brand and size of an individual product, or the brand, size, and quantity contained in cases shipped from manufacturers to retailers.

Product type data, obtained with the ease of a laser scanner, provides enough information to automate checkout lines or to improve inventory management. The Universal Product Code (UPC), established by the Uniform Code Council (UCC) in 1973, serves an important role in establishing uniformity and order concerning the product type data read from bar codes.[5] As of 2005, over 1,000,000 organizations use the EAN/UPC on their products.[6] This has unlocked enormous amounts of data to retailers and firms in other industries.[7]

With an increase in the intricacy and sophistication of products, the needs of business are moving beyond identification of product type to unique identification of individual objects by serial number. This represents a significant transition because unique identification introduces a much greater degree of complexity in system management.

Yet at the same time, unique identification also offers a number of possibilities that include greater visibility and real-time control of objects located anywhere between the manufacturer and the customer. The full realization of these capabilities will most certainly revolutionize the practice of supply chain management.

Organizing for Unique Identification

The EPCglobal Network is a system that builds on the tradition of automatic identification first established by the bar code. At the core of this system is the Electronic Product Code (EPC™), a serial numbering system designed to handle unique identification of trillions of objects. This numbering system forms a standard, uniform basis for linking physical objects together within a network that applies to all levels of the supply chain and across industries.

The means of creating such a network involves the placement of low cost Radio Frequency Identification (RFID) tags on objects such as cases, pallets, or individual products. In addition, RFID technology provides the capability for these tags to communicate with the Internet, secure Intranets, or point-to-point communication between organizations, through readers situated at various points within the supply chain. In the future, this type of network will form the base for "pervasive computing capabilities embedded in our everyday environments." [8]

Given such a computer infrastructure, it becomes possible to manufacture smart objects that can "sense," "do," and "understand." Resembling tiny robots, smart objects are things that can make decisions independently based on external data gathered through sensing technology combined with computer logic imbedded into the object. Examples might include consumer products that can automatically change price based on sensing supply and demand conditions on the retail shelf and the likelihood of going out of stock.

The foundation for creating smart objects is unique identification that only the EPCglobal Network can provide. In simple terms, the goal is to make new connections between physical objects regardless of location within the supply chain.

Positioned to be one of the major advances of the 21st century, this "web of things" is a significant innovation that has the potential to affect nearly all of commerce.

Creating a Global Standard

An important aspect of the EPCglobal Network is the development of common standards. Considered the bedrock of commerce, standards enable interoperability and the free flow of various business transactions. Often following technological breakthroughs, standards setting efforts have a positive record in driving economic growth that dates to the origins of trade in the ancient world.

During modern times, the co-development of new technologies along with the establishment of common standards represents a complex activity involving many different groups that must depend on each other for mutual success. Given expanding trade between countries, the interaction between new technologies and standards setting bodies will play an increasingly critical role in guiding the direction of innovation and future economic development.

As businesses begin the process of discerning the importance of the EPC-global Network to profits and revenue growth, it is essential to know the fundamental aspects of the technology and to be able to abstract these capabilities to the practical applications of the present and future.

The Basic Elements of Unique Identification

The original designers of the EPCglobal Network had a predetermined idea of how to use the Internet as a means of implementing unique identification.[9] Early researchers also introduced the goal of interoperability across all levels of the supply chain.

The basic design called for three components. First, and perhaps most important, a low cost RFID tag serves as the base of the system. To reduce the cost of the tag, researchers focused on new methods that would turn the manufacture of tags into a mass production process.

A second major development involved placing a unique serial number on the RFID tag capable of identifying trillions of objects. This task involved the calculation of the size of the number to ensure adequate coverage. The numbering system developed during this phase of research became the EPC.

The last step was to build a computer infrastructure capable of processing identification data and information. Since initial designs of RFID tags called for minimum functionality as a means of reducing costs, the EPC became the only data contained on tags. It served as a pointer to greater amounts of data and information held in a network. In this way, an efficient balance was struck between the limited functionality (and cost) of RFID tags, and the ability of computer networks to hold important data and information about objects.

With these components in place – low cost tags, the EPC, and a network infrastructure for handling data – the EPCglobal Network is capable of new types of data and information exchange as compared to current bar code systems. In business, it is a general rule that the value of information increases when it moves beyond the four walls of a firm. Many firms from a wide range of industries are convinced collaboration provides extensive benefits and are looking for new technologies to enhance its application with trading partners.

The greatest value of the EPCglobal Network may be the set of standards necessary for supply chain wide communication of unique identification data. In this way, collaboration between trading partners can take place on a more sophisticated level and it becomes theoretically possible to monitor and communicate with objects at any step of a supply chain. This type of collaboration and control has definite benefits that are quantifiable once the EPCglobal system is in place.

RFID versus the EPCglobal Network

The use of RFID tags to identify and track objects is not a new technology. The beginning of modern RFID traces to WWII where the military used transponders for the important purpose of identifying a returning aircraft as friend or foe.[10, 11]

Since the 1940s, businesses have applied RFID in a number of industry situations, mostly in the area of asset tracking, collection of highway tolls, security access, and consumer convenience. It is important to note that at least one inventor proposed the use of RFID during the 1970s as an alternative to the bar code.[12] At that time, RFID was about seven times more expensive than bar codes, prompting supermarkets to choose the latter.

Perhaps one of the most famous RFID applications to date is the ExxonMobile SpeedPass. With this service, consumers can purchase gasoline using a key chain fob that contains an RFID tag. When the customer arrives at a gasoline station, and begins pumping, an RFID reader located nearby automatically identifies the customer and charges their account upon completion of the transaction. This all happens without ever inserting a credit or debit card into the card-reading unit located at the pump. As a testament to its popularity, there are currently over 6 million users of ExxonMobil SpeedPass in the United States.[13]

A common characteristic of these early applications involves a "closed loop" approach. In this situation, RFID communication is tightly coupled and applications are highly specific. Proprietary standards dominate these systems and limited interoperability exists.

This is in contrast to the EPCglobal Network, where open standards enable supply chain wide interoperability. With this approach, there are many more opportunities to exchange data across multiple levels of a supply chain or across different industry sectors.

Charting the Future

There is no question that the EPCglobal Network and RFID technology have great potential to provide detailed data about objects within a supply chain. Current forecasts put the build-out of the technology on a gradual pace with the first comprehensive applications being in place by 2007.[14] As companies install dense ubiquitous reader networks within manufacturing facilities and supply chains, greater amounts of data will become available with improved accuracy and timeliness.

This represents a growth industry for vendors of RFID technology. ABI Research of Oyster Bay, NY estimates that RFID hardware and software sales alone have reached $1.53 billion in 2004, up from $915 million recorded in 2000.[15] Forecasts for the future size of the RFID industry vary a great deal with some predicting sales as high as $4.6 billion by 2007.[16]

Mandates in the retail industry[17] along with interest by the Department of Defense and the Food and Drug Administration will almost certainly ensure growing demand, although the near-term value to retailers and manufacturers remains unclear. If the past is any guide, one author predicts that in a high tech gold rush like RFID, it is usually those who sell the picks and shovels who make all of the money.[18] In this case, RFID tags, readers, and supporting software are the modern day equivalents of the picks and shovels.

The development of the EPCglobal Network and RFID technology will undoubtedly take many turns in practice. It is seldom that new technology finds application without a great deal of experimentation and a number of failures. However, the process of creative destruction, an economic principle first put forth by Joseph Schumpeter, ultimately means that mature technologies like the bar code will eventually make way for new innovations like the EPCglobal Network and RFID technology.[19] Though the bar code will be a mainstay of business for many years to come, its replacement is inevitable.

The task that business, academia, and non-profit standards organizations now face is the true measure of any innovation; initiating widespread application through a convincing argument of commercial value. Those who deal firsthand with the everyday problems of business will be instrumental in this effort.

Initial Application Ideas

During the early stages of development, practitioners and researchers have identified a number of applications that utilize the data and information anticipated from the EPCglobal Network. Some applications include track and trace within entire supply chains,[20, 21] theft detection,[22] improved service parts inventory management,[23] product obsolescence control,[24] and the management of production and logistics within military and civilian supply chains.[25] With almost all of these initial application ideas, the aim was to use information as a way to improve supply chain control.

In some cases, the EPCglobal Network holds the potential to go beyond information-based applications by changing basic business processes that have historically formed the structure of commerce.

For example, in the consumer goods industry the business model of manufacturers producing products that are sold to retailers, which in turn sell to customers, has remained essentially unchanged since the rise of department store chains during the late 19th century. With the supply chain visibility that the EPCglobal Network enables, it is possible that manufacturers can "own" products until the time of purchase by consumers in retail stores. Often called scan-based sales, this approach would redefine the consumer goods industry, turning retailers into providers of real estate for multiple stores within a store, and allowing manufacturers to make basic merchandising decisions involving price and promotion.[26]

This type of potential application highlights the inherent characteristic that networks often transcend the established boundaries of organizations.[27] Ocean shipping, roads, railroads, air transportation, telegraph, radio, television and the Internet are all examples of networks that contributed a great deal to economic growth, yet have the unique characteristic of cutting across significant business, governmental, social, and geographical boundaries.

As a final note, innovations like the EPCglobal Network also have a powerful transformative property. Some argue that networks often alter new inventions and ideas into something much more influential.[28] Steam power is one example. During the early stages of the Industrial Revolution the greatest changes did not come from the steam engine itself but rather from the invention of the railroad network, which utilized the steam engine to increase the average speed within a land-based transportation network.[29] Without the railroads, the agricultural and manufacturing capabilities of the 19th and 20th centuries would never have fully developed, and the United States would never have emerged as a world power.

While it is hard to predict what other technologies the EPCglobal network might transform, a precedent exists that such transformations will take place with the result of improved productivity.

The History of Technological Advances

Taking something new like the EPCglobal Network and RFID technology from an initial idea to wide-scale industrial application requires an appreciation of the history of technological advances. An understanding of this history will help to form the perspective that only the past can provide.

Seldom originating from a single source, innovations like the EPCglobal Network are often the result of many incremental improvements in technology that in total add up to big things. This type of process, which thrives in a free market economy, depends on communication and information sharing through various means, such as academic journals and the business literature, to stimulate ideas among individuals or small groups that possess a competitive desire to make something better. Historically, the result of such a process, as apparent with most of the significant mechanical inventions of ancient times, the middle ages, and the early stages of the Industrial Revolution, is that many innovations bear little resemblance to anything that had happened previously.

Archimedes' levers and pulleys, Gutenberg's printing press and movable type, Watt's steam engine, Whitney's cotton gin, and the Wright's airplane, were all innovations that changed the course of economies and history but had no prior equivalent. Though these engineers seldom created their devices without building upon the work of others, their unique insights about the potential of applied technology served to harness existing knowledge in new ways. Breakthroughs occurred primarily because of incremental innovations and the ability and focus of a single person to create something of economic value.

One invention alone, the printing press, accounted for massive increases in productivity and represented the first case where a mechanical device became an integral part of data and information creation though the manufacture of newspapers, pamphlets, and books. By a single account given in the late 18th century, a book would have cost 100 times the present price had the efficiencies of the printing press not existed.[30]

In situations where innovations are successful, it sometimes takes years to recognize applications. Inventive inspiration, often rooted in the idea of incremental improvement, sometimes focuses on the potential of what is, rather than the expectation of what might be. This means that inventors tend to overlook the practical applications of the things they create. Some researchers go as far as to assert that new technologies are not productive until nearly a generation after introduction.[31] The argument is that it takes time to learn how to use the technology in new ways and for new ideas to diffuse through the economy.

A case in point is the invention of air flight.[32] The Wright brothers first flew their machine on December 17, 1903, yet it was 1906 before issue of the first patent and 1909 before the US Army took delivery of the first aircraft.

Another important aspect of innovation is relevance along with the source of the initial inspiration for design. During the age of the Industrial Revolution, the great innovations in devices involving the manipulation of force and mechanical power frequently traced to a close relationship with the practical problems of everyday life. For example, the design of machines that replaced the backbreaking work of agriculture often mimicked the observed physical motions of humans working in the fields.

In perhaps one of the best descriptions of this type of innovation, Sherwood Anderson put forth a compelling fictional account of agricultural innovation in a small Midwestern town during the late 19th century. With a single sentence, he sums up the prevailing attitude of this time of great American innovation, "Do little things well and big opportunities are bound to come."[33]

Parallels with Identification Technology

The EPCglobal Network shares all of the characteristics of innovation that have previously occurred in the Industrial age. The technology carries the mark of its predecessor the bar code and the Uniform Product Code (UPC), first invented in 1948 (bar code) and fully implemented in 1974 (bar code and UPC).[34] Yet the idea of unique identification at the object level with RFID tags that communicate over a distance was beyond anything remotely considered by those who initiated the first bar codes.

Through the combination of RFID tags and a sophisticated information technology infrastructure, the EPCglobal Network has the potential to create a revolution in the quantity, quality, and timeliness of data generated within supply chains. In this way, the EPCglobal Network is a groundbreaking technology much like the printing press, which revolutionized communication, learning, and information sharing. Improving the quantity, quality, and timeliness of data has the potential to lead to better planning and control internal and external to business organizations, and the widespread application of automation.

Even though the EPCglobal Network has impressive potential, there also exists the risk of failure because of infrastructure complexity. To date, considerable theoretical development has taken place and prototype applications of the EPCglobal Network have proven successful in practice. However, the viability of large-scale implementation across a number of industrial sectors remains an unanswered question.

It is also true that other technologies might supplant the EPCglobal Network before full implementation. Though few such technologies are currently on the horizon, businesses in a free market economy have exceptional capability to identify and put into practice least cost alternatives that might include non-technological solutions to the various aspects of object identification within supply chains.

Finally, because it is a new technology, the most innovative applications of the EPCglobal Network are in the future and will probably have an origin in the everyday problems of commerce that cause loss of productivity. Much like the advent of air flight, it might take years to comprehend the full potential of the EPCglobal Network and RFID Technology. Taking the bar code as an example, more than 25 years passed between the invention of the bar code and widespread commercial application.

Productivity Through Information

The essential nature of the EPCglobal network differs from the devices developed during the age of industrial revolution in that the primary goal is production of data and information. Lacking the tangible aspects of mechanical innovation, this new technology depends on using data and information to enhance practical decision-making in business rather than a focus on improvements in equipment, training, or methodologies, which were the traditional ways of increasing manufacturing productivity.

In support of the productivity potential of the EPCglobal Network, evidence exists that the information economy has created significant results. According to a Federal Reserve report published in 2000, productivity growth has increased an average of one percent when comparing the first half (1990–1995) to the second half (1995–2000) of the 1990s.[35] Two-thirds of the increase has been traced to better use of information technology and increasingly efficient production of computers.[36] Further, a significant portion traces to better supply chain management in the retail industry, an outcome of increased data about operations.[37]

Conquering the dynamics of supply chains can only happen through the innovative use of information. As Harvard psychologist Donald Cox once stated, "information is the antidote to uncertainty."[38] Having real-time information on the location of an object within a supply chain improves visibility and enhances such critical business functions as inventory control and the delivery of customer service. To date, various information technologies including the bar code have already had an important impact on the

US economy by helping to mitigate inventory imbalances that often caused deep recessions.[39] The EPCglobal Network and RFID technology have every prospect to strengthen this trend.

CHAPTER 2

Hardware: RFID Tags and Readers

The EPCglobal Network and RFID technologies are much more than a replacement for the bar code. Designed as an information gathering system in addition to a means of automatic identification, the technologies establish a common base for the manufacturing and the supply chain systems of the future.[1]

The essential concept of RFID involves attaching a radio frequency identification (RFID) tag to a physical object. The tag either emits a signal continuously or it is activated to transmit information upon receiving a predetermined signal from another source.

RFID technology is often thought of as being similar to connecting a low cost, two-way communication device, like a cell phone, directly to an individual item, case, or pallet. Accomplishing this type of connection requires a great deal of engineering to develop cost effective systems that can handle large scale applications.[2]

RFID tags provide the capability for seamless and continuous two-way communication as an object moves through a supply chain. This means that any object bearing a tag can become networked without human intervention or manipulation by automated machines, as is the case with bar codes.

Most of the time, the tag is placed on the outside of a package, such as a retail item, a case, or a pallet of merchandise. In some situations, it is also possible to place the tag on the inside of packaging or as part of the product itself. For example, according to Dr. Daniel Engels of MIT, elecronic manufacturers have experimented with integrating tags directly into products such as stereos, digital cameras, or home entertainment centers.

A network of readers is the means used to automatically communicate with tagged physical objects that might be stationary or moving through a supply chain. Reader arrays have been fabricated and integrated in floor tiles, carpeting, shelf paper, cabinets, appliances, and shopping carts.[3, 4]

Generally, a reader has a specific field for detection of the tag that depends on the frequency of the signal. For a given power level, the higher the frequency, the greater the area that the field covers. In situations where there is little physical space, multiple antenna's can extend the effective range of a reader. When a tagged object enters the reader field, two way communication begins allowing for the exchange of information between the object and the reader. The type of information exhanged varies in complexity, ranging from a simple identification code like the EPC, to telemetry involving measurements of environmental parameters such as temperature or humidity within the proximity of the tagged object.

Tags come in many sizes and are durable, withstanding a number of different environments. The integrated circuit portion of the tag size is small, on the order of several millimeters in diameter. However, the antenna is much larger. The size depends of the frequency of the tag. Figure 2-1 shows a close-up picture of a tag, used in an early supply chain test. In this case the size is about 5 centimeters square with a thickness of about .2 mm.

Figure 2-1 – A Passive RFID Tag

Photograph Credit – the MIT Auto-ID Center

Figure 2-2 shows a picture of a reader (manufactured by Thingmagic). Resembling a metal plate, readers take up relatively little space and can be mounted in a number of different positions.

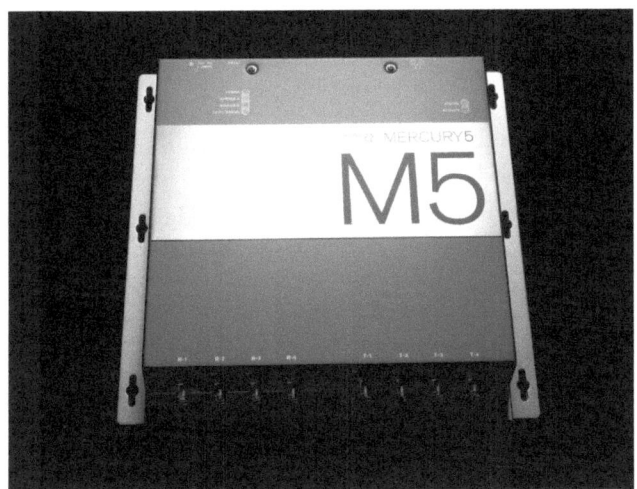

Figure 2-2 – A Reader

Photograph Credit – Thingmagic

Advantages of RFID Relative to Bar Codes

Few other inventions developed during the 20th century have had as wide an impact on everyday life as the bar code.[5] First implemented in 1974, the bar code has drastically reduced the amount of labor needed to operate retail stores, improved unit pricing and inventory accuracy, and shortened countless checkout lines.

Beyond retail stores, bar codes have been applied in many other situations to provide important information such as the coordination of production within manufacturing plants or tracking data for overnight packages in transit. Bar codes transmit a small amount of data that identifies the manufacturer and links to a description of the object. Non-profit standards groups such as GS1 and others administer the numbering system used for the bar code ensuring identification without duplication by other firms.

New research and development efforts have led to the development of the two-dimensional bar code that is able to carry more data about an object.

This opens possibilities to attach important information such as billing details directly to the object as it passes through the supply chain.

A basic characteristic of all bar codes is that a predetermined set of information travels with the object. In the case of a two-dimensional bar code, more information travels with the object as compared to a regular bar code. However, all bar codes have limitations including:

- The need for a direct line of sight from the scanner to the bar code,
- The ability to read only one code at time, and
- Bar codes often require human intervention to capture data or to orient packages.

In addition, bar codes provide only one-way communication and seldom provide real time information or Internet connectivity to the data. There is always a chance the bar code will be misread or in other cases, read twice. In addition, bar codes can be damaged or compromised in a way that makes them impossible to read.

RFID technology is designed to overcome all of these limitations and make it possible to automate the scanning process, providing real-time data within supply chains.

The Gradual Movement to Electronic Tags

In the last fifteen years, refinements in the design of integrated circuits along with advances in the manufacture of electronic tags have led to a decrease in cost per unit. Though the costs are still well above bar codes, the decreasing price of RFID tags opens the possibility for wider application in practice.

In addition to the advances in manufacturing technology for integrated circuits, there are several other important aspects worth noting that deal with the way tags are powered. Currently there are three basic types of tags used most often: 1. Active, 2. Passive, and 3. Semi-passive.

An *active* tag requires a small battery that provides electric power to continuously generate and transmit the radio frequency (RF) signal. Active tags can be read by readers (capable of receiving RF signals) located within the supply chain from a relatively long range – up to 30 meters. In general, these tags have significant amounts of memory to store information such as bill of lading details. In some cases, specialized readers called interrogators can read data from an active tag and send signals to reprogram the tag with new information or instructions.

However, active tags have several drawbacks, including signal interference, expense, and battery life. Because these tags transmit signals significant distances, there is greater chance of a "frequency collision" with other electromagnetic waves such as those emitted by radios, transformers, or cellular phones. This type of interference could cause the reader not to pick up the tag signal. Another issue is that with longer read distances, the opportunity of providing exact location information diminishes. Since one of the anticipated benefits of RFID and the EPCglobal Network is unique identification of an object and its location, active tags are not suitable for some applications where exact location is a critical piece of information.

In addition, the tiny batteries are somewhat expensive, limiting widespread use. Common prices for active tags range from $5 or more per unit, depending on capability, memory, and order size.[6] Because of relatively high prices, active tags are generally used for high value objects or in situations where the tag can be re-used a number of times.

Beyond the expense, the other disadvantage of active tags is that the batteries eventually wear out resulting in total loss of signal. This is disastrous if the tag fulfills a critical function such as tracking and tracing a physical object. Battery life varies a great deal depending on many different factors, so it is difficult to predict when a failure might occur. Though battery technology has improved for active tags, there will always exist a chance that battery wear-out will cause loss of signal.

Given the cost of active tags, industry and academics have undertaken research to develop low cost *passive* tags as an alternative.[7] With this technology, each tag does not contain a battery. Rather, the energy needed to power the tag is drawn from electromagnetic fields created by readers that also serve a dual purpose of gathering the signals emanating from the passive tags. The read distance of a passive tag is usually between five to seven meters.

Since no fixed power source is required, passive tags hold a great advantage over active tags in terms of lower cost per unit. This opens the possibility for the use of passive tags in a far greater number of applications.

Gradually, as costs decrease, passive tags will challenge bar codes as a means of gathering information within supply chains.

Designed to operate at low energy levels, passive tags store relatively little information. Just enough memory exists to store a serial number (the EPC). Information is stored on the Internet, or other private network, not on the tag. This provides a distributed means of holding information, which is very efficient.

A third type, the *semi-passive* tag, is a hybrid of both active and passive tags. It has a smaller battery that is partially recharged each time the tag enters the electromagnetic field of the reader. These tags are currently under commercial development and are not widely used in industrial applications though there is promise that such technologies might be an important enabler in the future. For specific applications, some semi-passive tags are just now coming into use.

In summary, there exists a range of choices concerning the type of RFID to employ within supply chains (see Table 2-1).

Table 2-1 – Comparison of Different Tags

	Active	**Passive**	**Semi-Passive**
Power Source	Battery	Induction from electromagic waves emitted by reader	Battery and Induction
Read Distance	Up to 30 meters	3–7 meters	Up to 30 meters
Proximity Information	Poor	Good	Poor
Frequency Collision	Hi	Medium	Hi
Information Storage	32 kb or more. Read/Write	2 kb Read only	32 kb or more. Read/Write
Cost/Tag	$5 – $100	10 ¢*8	Under Development, Some commercial applications

Because of lower cost, passive tags hold the promise of ubiquitous application to objects within a supply chain. The next section provides an overview of passive tags important to practitioners.

Technical Aspects of Tags

According to Professor Sanjay Sarma of MIT, a passive RFID tag has four components; the integrated circuit (IC), the antenna, the connection between the IC and the antenna, and the substrate on which the antenna resides.[9] In the past, the cost of tags relative to bar codes limited wide-scale application because traditional integrated circuit manufacturing techniques were not economical for production of large quantities of cheap tags.

New methods developed in academia and industry offer the prospect of continuous manufacturing of tags, a radical change from previous methods of producing integrated circuits. Using methods such as fluidic self assembly (FSA), originally developed at the University of California, Berkeley, companies such as Alien Technology are producing passive tags on rolls, similar to label production.[10] In addition, new ways of producing antennas using conductive ink printed directly on packaging such as corrugated boxes, will contribute to a goal of producing a 5-cent per unit tag.[11] While this cost still exceeds the cost of a bar code, it is anticipated that a 5-cent tag will be economically viable for use with cases and pallets.

Besides mass-producing passive RFID tags, technology vendors are also focusing on reducing the cost of readers to the range of $100 to $200 per unit.[12] In addition, several companies have developed special antennas that extend the capabilities of readers.[13] With multiple antennas attached to a single reader, the effective scanning range can be increased to 10 meters or more depending on the number of external antennas. This reduces the number of readers needed to cover an area.

Both passive and active tags operate at a range of different frequencies. Although there are few universal global standards, some common frequencies do exist. Examples include 13.56 MHz and 2.45 GHz, which are available in most parts of the world.[14] It is important to note that the 433 MHz frequency is used predominately for active tags.[15]

The frequency available for passive tags in the United States is the UHF spectrum between 902 and 928 MHz.[16] This is the frequency that the FCC has allotted as the ISM (industrial, scientific, and medical) band. Passive tags use the 915 MHz frequency, the midpoint of the ISM band.

However, this frequency is not universally available worldwide. For example in Europe, because of limitations on frequency allotment, the bandwidth is between 865.7 and 867.6 MHz. This provides a narrow two MHz bandwidth for RFID tags. China has not yet allocated a frequency range for passive tags, although it is expected that by 2007 an allocation will be made.

Because there is a desire to make RFID a global standard, the lack of a universally available frequency presents a number of difficulties in application. Some hardware vendors have developed an "agile" reader capable of working in a band of frequencies around 915 MHz.[17] This would allow for interoperability regardless of the country regulations that govern the exact frequency available. In addition, other developments in technology involving tag protocols also hold much promise in overcoming poblems associated with differences in frequency.

It also is important to understand that each frequency has strengths and weaknesses. For example, in general, passive tags that use a lower frequency experience a better scan rate when near metals or liquids. However, it is also true that lower frequencies mean a slower rate of data transmission. This limits applications where large amounts of data are being transferred.[18]

As a recent development (2006), at least one tag manufacturer has presented a strong case that tags using the 915 MHz frequency, combined with new tag standards mentioned at the conclusion of this chapter, perform very well when scanning occurs near metal or liquids (see http://www.impinj.com/). Practitioners should keep in mind that tag technology is evolving very quickly and new developments are being announced on a daily basis.

A final aspect of tags that is worth noting involves functionality. Keeping in mind that tags are like small computers, there are opportunities for different levels of sophistication. While passive tags generally have limited functionality, there are other classes of tags that have much more capability. Figure 2-3 [19] gives a rough guide to functionality.

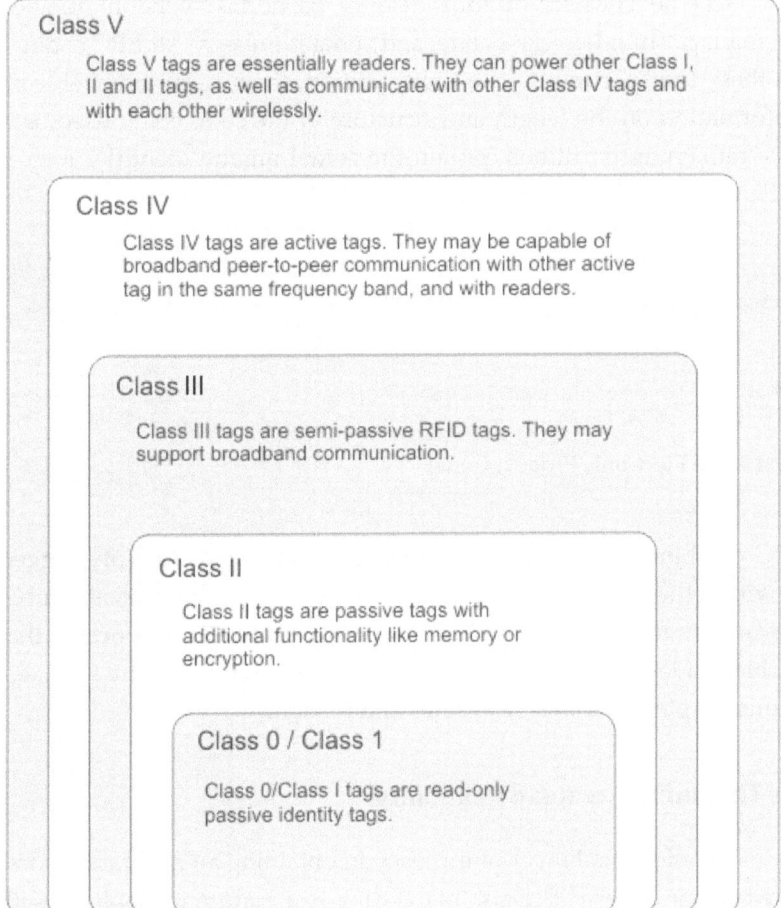

Figure 2-3 – Classes of Tags

The Electronic Product Code

The EPC is a numbering system designed to identify all physical objects and aggregations of objects.[20, 21] The EPC code is sufficiently large to accommodate all current and future naming methods. It provides for industry coding standards, such as those from GS1. These standards include the original Uniform Product Code (UPC), as well as other numbering schemes. The EPC is intended to be globally accepted. Since the EPC is used primarily to link physical objects to the network, it is designed to serve as an efficient information reference. A final point, the code is extensible, allowing future expansion in both size and design.

The EPC scheme consists of four distinct partitions: version number, domain manager number, class code, and serial number.[22] All EPC codings contain these four partitions. The first partition, the version number, contains information on the length and structure of the code being used, and the three remaining partitions contain the actual unique identifier for the object (Figure 2-4 shows the structure of the EPC).

Figure 2-4 – The Electronic Product Code

The EPC combined with RFID provides great power to identify tagged objects within the supply chain. Though unique identity enables a number of innovative supply chain applications, it is especially important that RFID achieve a level of reliability in order to gain credibility. The final section of this chapter examines read rates and reliability.

Factors That Influence Read Reliability

RFID adds an additional layer of intricacy in obtaining an accurate read as compared to bar codes.[23] Because bar coding is a mature technology with fifty years of testing and development, conditions necessary for successful production and use are well understood. Further, because bar codes depend on optical means for a successful read, the technology is direct and understandable. As long as the correct conditions exist, read reliability should be high.

Yet with RFID, tags are coupled to readers via radio-frequency fields and waves that are invisible to the human eye. As a result, read performance can seem highly variable and sometimes difficult to predict because it is hard to visualize the properties of electromagnetic fields. In addition, environmental factors play a much larger role in negatively affecting performance as compared to bar codes.[24] Materials surrounding or blocking tags, such as liquids and metals, can absorb and reflect the radio frequency energy. Humidity, not a factor in bar code reading, can significantly reduce the read range for RFID tags. A final complicating factor is that the manufacturing process for tags has still not achieved critical mass. In some cases, manufacturing imperfections lead to poor read reliability. This type of failure is independent of environmental factors influencing electromagnetic fields, and causes complexity in achieving high reliability.

Reliable and accurate reading of RFID tags is generally not a problem for a specific process if thorough testing and debugging is possible as part of the installation. However, in open system applications such as tracking an object throughout the supply chain, neither the applicator of a tag, nor the integrator of a reader installation have direct control over a single implementation. In this case, the proprietary model of deploying RFID is no longer viable because interoperability is not possible.[25] Current research and development efforts are focusing on standardization and testing to improve tag and reader designs so that true interoperability exists within supply chains.

Along the path toward 100% read reliability, many companies are considering adopting an "inferred read" approach. By associating all items within a case to that case, or all cases on a pallet to that pallet, a successful read of some fraction of the aggregation can be used to represent a successful read of all objects in the aggregation. For example, if a full pallet contains 60 cases (each tagged), then a successful read of only one of the EPC tags implies that a complete pallet has been read. The RFID approach, where information is held on the network rather than in tags, is a great advantage in facilitating inferred reads.

However, the inferred reads approach assumes that the aggregation is always intact (i.e., all items are in a case or all cases are on a pallet). This is a disadvantage when data is needed for such things as a drug pedigree where the EPC code for each package must be linked to previous shipments. If a high reliability read of RFID tags placed on each package of drugs is not possible, the tracing information must be entered manually. With the large volume of drugs moving through the supply chain, even partial manual entry of information needed for the pedigree might be overwhelming.

The Way Forward

By one estimate, "companies worldwide are expected to use more than 20 billion RFID tags and labels" by 2008.[26] However, before widespread application of RFID can occur within supply chains, tag and reader costs must decrease and read-rates must increase to the reliability demonstrated by the bar code. It is especially important that tag cost drop to single digits.[27] Given the perceived value of the EPCglobal Network and RFID technology, there are many efforts within industry and academia to make incremental engineering and production improvements that will establish RFID as the predominant identification technology for physical objects.

In December 2004, the standards for generation 2 (Gen 2) tags were established with hopes of improving interoperability between different manufacturers and reducing cost.[28, 29] Attracting a great deal of attention, the Gen 2 standard will also increase the speed and reliability of reading Class 0 and Class 1 tags. Though additional changes to standards are certain in the years to come, Gen 2 represents major progress in achieving the reliability, security, and speed needed for a "web of things."

Beyond standards development, there are substantial research and testing efforts underway to solve the practical problems of implementing RFID with a wide range of industrial environments. In 2004, the University of Kansas established the RFID Alliance Lab, an independent effort to conduct testing of tags and readers.[30] As part of an academic institution, the lab has no commercial interests that might influence testing. The work of the lab concentrates on intensive empirical testing and reporting of results.

Also in 2004, MIT established the Packaging Special Interest Group (SIG) as part of Auto-ID Labs.[31] With the purpose of investigating the impact of various packaging materials on RFID systems, the group concentrated on four areas:

- Design of a Field Probe
  Since the electromagnetic fields used as part of RFID are invisible to the human eye, tools are needed to measure the strength of the field and map its boundaries. Such tools are important in providing specific data for implementations of readers in industrial environments.

- Graphical Simulator
  Using the data produced by the field probe, the specific shape of the electromagnetic field can be visualized by a simulator. This type of visualization is a further aid to determining the optimal reader placement.
- Antenna
  Design of antennas represents an area for improved read ranges.
- Electromagnetic (EM) Propagation in RFID Systems
  A final area of the SIG involved understanding of how electromagnetic waves propagate between layers of cases on pallets and through various materials.

Receiving support from a consortium of eight companies, the Packaging SIG was successful in generating new research of value to industry.[32, 33, 34] This builds on earlier research that established a base for the study of open standards for RFID.[35]

Perhaps the greatest value of this type of industrial research effort in a university setting is the opportunity to stimulate innovations among each successive group of undergraduate and graduate engineering students. This building process is fundamental to the advancement of science, engineering, and industry. The research of the Packaging SIG has influenced other researchers working on the problem of increasing read rates.[36, 37]

In one case, a novel approach has been developed that involves a three dimensional "corner tag" with an antenna capable of receiving electromagnetic waves in all directions as compared to conventional planar tag antennas that are orientation-limited.[38] Called the Albano tag, it fits on the corner of a case. Preliminary testing shows that this approach adds little cost to applying the tag and improves the consistency of read rates in static situations.

As a final note, the long-term future might not depend on the manufacture of integrated circuits as an integral part of RFID tags. New technologies currently under development allow for printable electronics.[39] This raises the prospect of printing tags directly onto cases or individual items, reducing costs as compared to conventional tags. Though this technology is in the development stage, the flexibility of printable electronics at the point of application raises a number of new possibilities for the wide-scale deployment of RFID within industry.

While RFID tags and readers provide a great deal of opportunity to gather data from the supply chain, there remains the question of the infrastructure needed to organize data obtained from RFID tags. The next chapter examines infrastructure issues that are important to practitioners.

CHAPTER 3

Infrastructure: EPCglobal Network

Though tags and readers are important in achieving RFID on a large scale, there are other critical system elements needed for successful implementation in practice. The fundamental assumption of the EPCglobal Network is that low cost tags applied to objects will hold just enough data for identification using a serial number (the EPC). Additional information about a tagged object resides not on the object itself, but on a computer network. With this architecture, the serial number is the key for accessing information about the object.

For RFID Technology to become viable in practice, an infrastructure must exist for processing and communicating EPC data.[1] In meeting the goal of creating a common infrastructure, MIT announced Auto-ID Release 1.0 in October 2003. At the same time, MIT entered into an exclusive licensing agreement with GS1.

In turn, GS1 established a new division called EPCglobal to implement Release 1.0 and to conduct further development based on industry input. This put forth an initial set of standards that formed the basis of an infrastructure for EPC data. Later, Auto-ID Release 1.0 became the starting-point for the EPCglobal Network.

Representing a mature set of standards,[2] the original infrastructure design for linking physical objects to the Internet closely resembled that of the Internet itself. Distributed processing and open standards were the defining characteristics that combined to make the EPC and RFID technology operable across business and international boundaries. In the future, knowledge of this type of infrastructure will be as common as that of microcomputers, networks, and the Internet. All in business will need to understand at least the conceptual aspects of how the technology works in practice.

The EPCglobal Network and RFID technology have potential to combine the strengths of wireless broadcast networks such as television and radio, with the power of instant two-way communication.[3] This accomplishes a task of great value to commerce through the merging of information with physical objects. In essence, the EPCglobal Network creates an object-centric system based on unique identification. This type of infrastructure serves as a base for creating new forms of automation and ubiquitous computing needed for "smart objects" that will populate the supply chains of the future.

The next section explores the essential aspects of Release 1.0. This serves as an important base for understanding the current development of the EPCglobal Network.

An Overview of Release 1.0

In conjunction with advances by tag and equipment manufacturers, the objective of Release 1.0 was to establish infrastructure and set open standards for wide-scale adoption of passive RFID technology across many different industries, thus creating a web of things.[4] Encompassing a comprehensive information technology infrastructure along with open standards for data transmition, Release 1.0 marked a change in approach as compared to traditional RFID systems that depended upon proprietary standards. Though effective in specific applications, traditional RFID technology stopped short of completely fulfilling the need for open communication within business, government, and medical organizations.

As is the case with most information technology, the lack of open standards for RFID tended to inhibit widespread adoption within supply chains for consumer goods and other manufactured products because organizations were wary of implementing a system that might cause internal and external compatibility challenges. Interoperability between tags and readers situated at various stages of the supply chain was a major concern. This fact, combined with significant tag and reader costs that were dependent on volume of production, has limited widespread adoption of RFID in spite of advances in sophistication.

Beginning in 1999, the MIT research and development effort that culminated in Release 1.0 sought to overcome these limitations by taking advantage of the Internet as a blueprint for a future design that was not dependent on traditional RFID technology used in the past. The result was a comprehensive system that featured open standards and a flexible information technology infrastructure able to meet the needs of various entities within supply chains. From a practical standpoint, this meant that a tag produced by one manufacturer could be read using equipment produced by a different manufacturer. Interoperability between tags and readers is an essential condition for the wide-scale application of RFID technology.

The original design of Release 1.0 consisted of four components:

- EPC (electronic product code)
- ONS (object naming service)
- PML (physical markup language)
- Savant (data handling)

As mentioned in Chapter 2, the EPC is a numbering system that contains enough combinations to identify trillions of objects. This is necessary because the ultimate goal is to provide a structure for low cost identification at the individual item level, meaning every single object will have its unique code. From the early stages of conceptual development, researchers envisioned that tags would contain the EPC allowing for capture through RFID.

PML served as the communication format for the data gathered from tags.[5] Based on XML (extensible markup language), a computer language that has gained popularity in eCommerce transactions, PML represented a formalized data organization to store information in an open format that allows portability within organizations and externally among partners within entire supply chains. By having a standardized means of describing physical objects and processes, PML was designed to facilitate inter- and intra-company commercial transactions and data transfer.

The ONS acted as a pointer (or registry) to connect an EPC to a PML file stored either on a local area network or over the Internet. The ONS performs a similar function to the Domain Naming Service (DNS) of the Internet, which connects a text web address to an underlying IP address. An IP address is comprised of a 32-bit numeric address written as four numbers separated by periods, to find resources over the Internet. However, with the EPC, the ONS conducts a look-up on a serial number to find the product information linked to that number. This is the opposite functionality of the DNS, which uses text to find an IP address.

The original founders of the Auto-ID Center thought of the Savant as a "middleware" application that processes the data, and performs error checking and de-duplication procedures in the event that more than one reader receives a signal from the same tag. Primarily, the Savant was designed to deal with collection of local data from the readers. It handles the scalability problem associated with the massive amount of data captured by RFID through filtering and aggregation prior to transfer to the Enterprise applications.

In summary, the EPC identifies the product, PML describes the product, and ONS links them together. This forms a cohesive system with the flexibility to span industries and international boundaries, just like the Internet.

To make the system work, products are tagged with passive RFID chips containing the EPC. The tags are placed on surface areas of pallets, cartons or contained within item packaging. Readers are positioned at strategic points throughout the supply chain where companies need to capture data. Readers constantly emit an electromagnetic field that is received by the tags through a small antenna. This energy activates the tag, and in turn, a signal is generated and transmitted to the reader.

Through this process, readers capture the EPC and interact with a Savant to look up the information on the product using ONS. The position of the reader receiving the EPC signal provides important information on location, and environmental conditions such as temperature, vibration, and humidity, which is then linked through databases to the EPC. All this information is housed and written to corporate databases using the PML format. (See figure 3-1 – Technology Overview)

Testing of the technology associated with Release 1.0 began in 2001 and involved a complete supply chain comprised of suppliers, retail outlets, and technology vendors.[6] Companies participating in the study included Coca-Cola, Gillette, International Paper, Johnson & Johnson, Kraft Foods, Procter & Gamble, Unilever, Wal-Mart, Sun Microsystems, CHEP, and others.

Initial results of the study indicated that Release 1.0 was technologically feasible in practice. An important aspect of the study involved the use of a

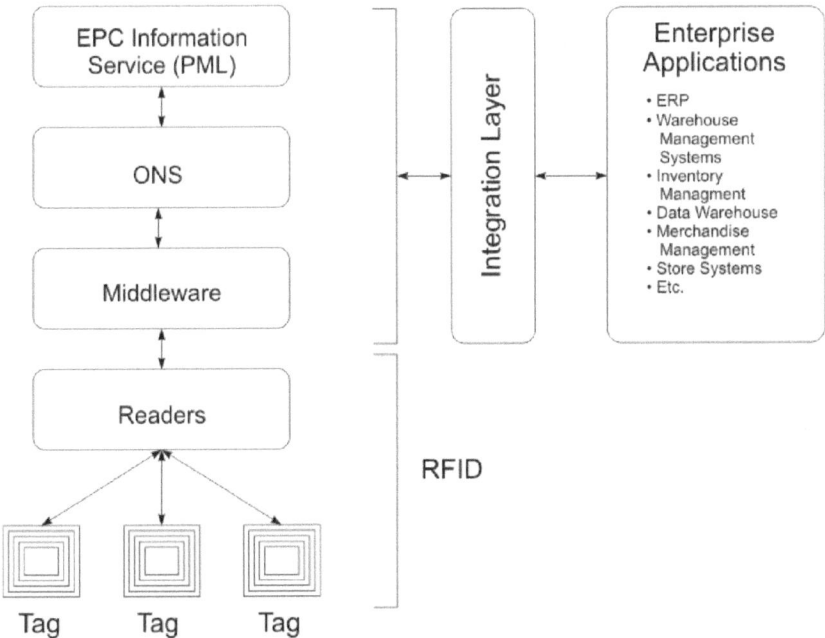

Figure 3-1 – Technology Overview

portal approach for reading tagged pallets. With readers placed at the entrances and exits of warehouses, 97% of all tagged pallets loaded with untagged finished goods were identified.[7] Though this was a tightly controlled test, the ability to identify pallets of finished goods using RFID technology and the EPC provided encouragement that with continuing refinement, Release 1.0 was viable in a real supply chain situation.

Developments Since 2003

Release 1.0 represented a major advancement in establishing interoperable systems for RFID using passive tags. However, the practical realities of implementation forced a re-thinking of the network architecture. In all subsequent updates, the EPC has maintained its role as the centerpiece of the system. However, the other components have undergone slight modification to meet the needs of industry.

From its inception in 2004, EPCglobal has grown to include over 800 member companies with 62% from North America and 20% from Asia (for more information, http://www.epcglobalinc.org/).[8] A large number of companies belonging to EPCglobal are from the food and consumer goods segments although other firms from industries such as healthcare, technology, and aerospace are also members.

With equal representation from both manufacturers of goods and technology vendors of tags, readers, and software, a substantial international RFID community exists. Given this base of interested parties, EPCglobal has established a number of working committees to adapt Release 1.0 to the needs of industrial supply chains. Under the leadership of the Business, Technology, and Public Policy steering committees, over 40 sub groups engage in regular meetings to perform the work needed for finalizing all standards associated with EPCglobal and RFID technology.[9]

These standards should be complete sometime in 2006.[10] Many believe the tipping point for implementation of the EPCglobal Network and RFID technology will come in late 2006 or early 2007 as the standards for the information technology infrastructure become set and the production and use of Gen 2 tags increases.[11]

Strengthening Authentication

As the various working groups of EPCglobal analyzed the implications of Release 1.0, attention began to focus on issues of security and authorized users. Keeping in mind that the original design of Release 1.0 was fashioned based on the Internet, attention focused on the centralized look-up approach for EPC data, which resembled the DNS. Though this approach would efficiently meet the needs of communicating the information associated with billions of EPC's being read in supply chains, there were concerns among various working groups within EPCglobal about what organization would control and administer the centralized look-up authority, namely the ONS.

While the DNS currently used for linking domain names to IP addresses is a fundamental component of the Internet, it provides access to public information contained in web pages written in HTML. On the other hand, the ONS as originally designed, links the EPC to data contained in private databases located within firms.[12] Using this approach eliminates the inefficiencies of point-to-point communication commonly associated with computer communication technologies such as Electronic Data Interchange (EDI).

However, the ONS also presents a significant security concern because it plays a vital role in connecting potentially confidential information about products like pricing or volume sold through a particular retail outlet. Further, if the customer information obtained at retail outlets becomes associated with a particular EPC, there are also concerns regarding privacy.

It is also true that in some cases, firms might not want to share certain types of product data with other trading partners.[13] Without modifications to the existing ONS, situations might arise where an EPC query by a trading partner such as a retail store might show that a product came from a particular manufacturer, but no other information would be available because the manufacturer had previously decided to keep all information contained within internal computing systems private.[14] This has critical implications for various supply chain wide functions such as inventory management, track and trace, and authentication, all potential benefits of the EPCglobal Network and RFID technology.

Though encryption technologies exist to protect the ONS from intrusion, several leading companies have expressed a preference to communicate EPCs and associated data by using direct database-to-database communication along with user authorization, thus bypassing the need for PML and ONS as envisioned in Release 1.0.

The term EPC-IS describes this new approach, which essentially is an interface for servers that store physical object information.[15] This approach considers not only the types of data associated with an EPC, but also the business processes that are a part of computer-to-computer communication. While it remains uncertain what final structure the architecture will take, it is clear that the basic functionality of Release 1.0 will continue into the future, although there will probably be more emphasis on private, security enabled, one-to one transactions involving the communication of EPC related data. There are indications that at least some members of EPCglobal are supporting the use of Applicability Statement 2 (AS2), which is "a method of doing electronic data interchange (EDI) over the Internet"[16] as an alternative to network-based management of EPC data.

A New Role for Middleware

The ongoing work of the EPCglobal committees along with initiatives in the vendor community and at universities, have also defined a new role for the EPC middleware, formerly called the Savant. In essence, the middleware is a layer that "filters data from RFID hardware devices and sends it to enterprise systems."[17] A new specification for application-level events (ALE) will link the software used to gather EPC data from readers with the EPC-IS that stores product level information.[18] All members of the EPCglobal Network will use the ALE.

Though technology vendors offer a number of different middleware solutions on a proprietary basis, there are also movements to make middleware an open system. Most notable is the work of the Radioactive Project, and free access to EPC middleware located at Sourceforce.net.

At least one university has moved into the area of developing middleware. Using a consortium approach that includes Intel, Northrop Grumman, and Hewlett Packard among others, the University of California at Los angles (UCLA) has been working in the area of RFID since 2002.[20] In 2004, the research group demonstrated middleware that runs on "a distributed system based on Web services and built on a Microsoft .Net framework." [21]

Since every firm that becomes a member of the EPCglobal Network will need to use some type of middleware based on the ALE standards, the market for software vendors is substantial. Leading companies such as OAT Systems, H-P, BEA, and SAP have participated in middleware development efforts since the early stages of the MIT Auto-ID Center. Through significant real world experience, these firms are on the cutting edge of building the systems needed to link EPC data gathered from readers to the complex Enterprise Resource Planning (ERP) systems in place within companies.

The EPC Simulator

An additional development worth noting involves an initiative begun in August 2005 to build a software simulation for studying the impact of the EPC on data flows internally within firms and across entire supply chains.[22] Led by MIT Auto-ID Labs, the simulation will be international in scope, involving other members of Auto-ID Labs located in Europe and Asia. The plan is to utilize simulated RFID read events associated with manufacturing facilities, warehouses, regional distribution centers, and retail stores. Objectives of this project are to test the viability of EPC-IS on a large scale and to help companies learn about the intricacies of handling EPC data in a simulated manner rather than through live experiments in operational supply chains. A focus will be on the issues of security and system capacity.

The anticipated completion of the project, which has the endorsement of EPCglobal, will provide "a reference implementation for software vendors that want to develop EPCglobal Network applications, and for end user companies that want to model their own systems before deploying EPC technologies."[23] This type of work is fundamental to the future deployment of the EPCglobal Network.

Kevin Aston, the V.P. of marketing at ThingMagic and former Executive Director of MIT Auto-ID Center stated at Frontline 2004 that "once the cost of the infrastructure is covered, the cost of the information is free." Like bridges, public utilities, and the national interstate highway system, investments in infrastructure are often significant and require intensive engineering to determine the optimal design. However, once standards are tested and set, designs can be duplicated many times, reducing the incremental cost of application and building a sustainable infrastructure that can return many future benefits.

The EPC Network follows this pattern and shares the characteristics of physical infrastructure development. While the EPC is in the beginning stages of application, there is no question that once the proper infrastructure is developed, benefits will flow to firms many years into the future. Already, the food and consumer goods industry has invested about $100 million in testing for 2005 alone.[24] However, for the technology to move forward, other industries must follow the same example.

Historical Reference

As a concluding note to this chapter, it is useful to provide some background concerning early technological innovations that influenced the development of the EPC.

Usually with most technology, the path to commercial application has its beginnings in loosely related research that offers an important fundamental insight. According to Dr. David L. Brock, a cofounder and former Director of the MIT Auto-ID Center, the roots of the EPCglobal Network trace to the idea of using the Internet as a means of controlling machines such as robots.[25, 26] Commonly termed "Internet devices" the first was the Trojan

Room Coffee Machine, which researchers implemented in 1992 at the University of Cambridge.[27] In this early example, a video card captured an image of a coffee pot every second and made this data available on a network, allowing researchers to see the status of the coffee without an unnecessary trip to the machine.

At about the same time engineers were exploring the use of Internet devices, computer scientists were working on a parallel track involving the development of standards for improved internet communication. The first was the Gopher protocol, introduced in 1991. Gopher grew rapidly as academics placed news and research information on the network. However, by the mid 1990s, Gopher had nearly disappeared. It was replaced by the World Wide Web, which featured an improved user interface and a free-form markup language.

Building on these advances in engineering and computer science, Dr. Brock founded the "Virtual Worlds Project" at the MIT Artificial Intelligence Lab in 1994. This was an early effort to link networked devices, such as robots and home appliances, with mathematical models located on a network. The models provided various instructions to operate theses devices automatically.[28, 29, 30, 31]

With these improvements in Internet control there remained a single, fundamental problem, the lack of unique identification for physical objects. Initial research efforts experimented with bar codes as a means of object identification. These efforts failed because bar codes required object manipulation to acquire a clear line-of-sight for a scan. At the time, the bar code also did not provide the unique identification needed for automated processes.

To solve this problem, Dr. Brock acquired some radio frequency security tags from a small company named Arizona Microchip. These tags operated at low frequency (135 kHz) and had 64-bits of raw data storage. The idea was to store all relevant data on a tag affixed to an object. Using a reader, the tagged object could be scanned to retrieve the data.

With this approach, it quickly became clear that the amount of data associated with each object far exceeded the amount of storage available on the tag. The only viable alternative involved storing a minimum amount of data on the tag consisting of some sort of identifier, which could then be linked to larger amounts of data located on the Internet. This approach required the development of a globally unique identifier for RFID tags.

After initial contact with Thomas Brady, a senior scientist at the Uniform Code Council (UCC), it became clear that others were working on similar ideas involving electronic identification. Viewed as a potential replacement for the bar code, an initial research effort was formed at MIT called the Distributed Intelligent Systems Center (DISC) to develop unique identification methods needed for the automation of physical objects. Professor Sanjay Sarma, who held an assistant professorship appointment in the MIT Department of Mechanical Engineering, joined Dr. Brock and Professor Sunny Siu in this initial effort. Kevin Ashton, who worked at Proctor & Gamble and was thinking about applications of RFID to reduce stock-outs on retail shelves, met Brock, Sarma, and Siu during this time.

About one year later, this led to the establishment of the MIT Auto-ID Center (1999) that focused on supply chain applications of unique identification. Alan Hamerman, who had previously played a pivotal role in the development of the bar code in the consumer goods industry, became interested and provided the necessary credibility and contacts to help establish the MIT Auto-ID Center as an industry sponsored research project with broad support.[32]

In summary, it was through an interdisciplinary approach combined with various forms of communication, sometimes by chance, that several different parties came to the understanding that ideas intended for the improvement of automation and robotics, namely the use of RFID and unique identification, also applied to the identification of objects within the supply chain. This has led to the notion that physical objects could have the capability of "self identification."

Taking this a step further, Chapter 16 examines the idea of self-identification as applied to non-physical objects, namely data, information, and mathematical models. Like the EPCglobal Network, this new approach has potential to create significant improvements in productivity.

CHAPTER 4

Data: What, When, and Where?

As companies deploy the first generation of RFID systems based on the EPCglobal standards, complex issues are arising involving the end use of the data. One area of immediate interest involves the type of data to capture from the RFID read event. In simple terms, this means the what, when, and where for a RFID tag affixed to an object. Decisions (and potential standards) for data capture must take into account the uses of the data in a host of computer applications, including the need to achieve supply chain wide visibility.

During the implementation of RFID technology, questions continually arise concerning the time and location of the read event associated with an EPC, and the sharing of this data between trading partners. Since the installation of all RFID middleware requires the identification of reader locations, there are underlying operational issues concerning a universal standard for location.

Choices of what data to capture during the read event also impacts the interface with legacy information networks and systems, such as Electronic Data Interchange (EDI), Warehouse Management Systems (WMS), Transportation Management Systems (TMS), the Global Data Synchronization network (GDS™), and Enterprise Resource Planning (ERP) systems. These systems serve as the bedrock of commerce for a wide range of industrial firms.[1]

An important characteristic of future supply chains will be the ability to locate objects that might be stationary or in transit, potentially anywhere within or beyond the boundaries of a company. Having a common set of specifications for location is essential for various supply chain functions such as track and trace.

This chapter explores the issues associated with defining location in a robust way to meet the needs of supply chain stakeholders. The lack of a standard framework for location is a major obstacle that impedes interoperability. If non-standard definitions of location proliferate, it will become increasingly difficult to achieve data integration across supply chains that might span several different industries.

In the next section, a simple example is put forth to demonstrate the complexity involved in defining location within RFID systems. Later sections deal with the importance of location in day-to day commerce and a discussion of the approach proposed by EPCglobal.

An Example from the Consumer Goods Industry

The concept of a precise location is a relatively recent development in the course of history. For example, a letter sent in London during the late 1600s might include the following address:

> To Mr. Henry Olenburge at his house about the middle of the old Pal-mail in St. James Fields in Westminster [2]

Though modern addresses have greater consistency, there continues to be significant international interoperability issues as countries choose different ways to express location for businesses and residences.

The complexity in designing a standard, machine understandable code structure for location arises because in almost all cases the locations of an object are relative to something else. Since commerce is composed of a large number of objects, fixed and moving, location has a vast number of answers depending upon the frame of reference. Further complicating matters, location is described with words that can have multiple meanings depending on the context of usage. No single semantic definition exists for words used to describe location.

An example of this complexity is the simple question "Where is my shipment?" Does this mean the current in-transit location relative to a roadmap? Or, where is the shipment located on the surface of the earth – latitude, longitude, and altitude? Alternatively, where is the shipment inside its shipping container?

The answer to the question "Where is it?" likely contains several parts. For example, suppose a supply chain manager asks, "Where is my shipment of Proctor & Gamble (P&G) Bounty Paper Towels?"

The answer might be "the shipment is in the Proctor and Gamble Manufacturing Center, Cape Girardeau, MO, USA."

This phrase contains the following parts: (1) my shipment, (2) Proctor and Gamble Manufacturing Center, (3) Cape Girardeau, (4) MO (Missouri), and (5) USA. Each of these parts represents a progressively larger spatial area. Therefore, a simple expression for machine understandable location might be:

P&G Manufacturing Center,

in Cape Girardeau, in Missouri,

in the United States of America.

For this example, the street address is sufficient to identify location. The exact location of Cape Girardeau, in terms of longitude and latitude, within Missouri is not an important information element for many transactions.

However, RFID tags containing the EPC have the potential for more detailed information about location. Using another related example, a supply chain manager might ask a question about the location of a particular item, P&G Ultra Downy Mountain Spring, and receive the following response:

P&G Ultra Downy Mountain Spring,

item epc:id:gid:37000.35830.344098943

is located in **case** urn:epc:id:sgtin:37000.800031.400

is located on **pallet** urn:epc:id:sgtin:0652642.800031.400

is located on the **top shelf of Shelving Unit C**

Proctor and Gamble Manufacturing Center

Cape Girardeau, MO

USA

From this example, a comprehensive solution to the RFID location issue contains a number of different elements: (1) street address, (2) logistics containment (case and pallet), (3) spatial location (inside a warehouse), and (4) geo-position (longitude and latitude). Connecting these elements in a standard that is portable across different industries is a challenging task. However, achieving a common representation for location is a fundamental building block in gaining supply chain visibility along with other goals such as automation.

Why Location Is Important to Business

Increasingly, governmental agencies are holding manufacturers accountable for goods that flow through distribution partners to the end consumer. This might include pedigree tracking as a means of reducing counterfeit, or providing accurate records for taxation purposes. In both cases, a common definition of location becomes essential to generating a robust tracking and tracing capability.

For example, the Food and Drug Administration (FDA) has announced a guideline that recommends tracking and tracing of drugs through complex supply chains by using unique identification technologies.[3] Given location data about a drug as it passes through the supply chain, a pedigree could be generated showing the entire history of movement and the time spent at each location. If pedigrees accompany shipment through the supply chain, it is less likely counterfeit will occur because buyers will not purchase pharmaceuticals with a suspicious history of movement.

The European Union provides another example of imposed requirements for track and trace. In this case, the initiative involves verification of Value Added Tax (VAT) collection. In July 2004, the world's top tobacco company, Philip Morris, agreed to pay $1.25 billion to the European Union over 12 years to fight contraband cigarettes and to end legal disputes with the EU over smuggling charges. Regulators in Europe have been holding manufacturers (in this case Altria Group, Parent Company of Philip Morris International) responsible for control of supply chains from start to finish, regardless who puts its products in the hands of consumers.[4]

In both situations involving drug pedigrees or VAT tax compliance, the need exists for a standard definition of location. Yet there are few location standards in place for specific industries. In no case is there a universal standard that encompasses all of industry.[5]

Existing Location Standards

The Global Location Number (GLN) is the standard for data shared across the Global Data Synchronizations Network (GDSN™), operated by GS1 the parent organization of EPCglobal. The GLN is simply a thirteen-digit number associated with a static location that could be in a manufacturing facility or customer distribution center, or the back room of a retail store. The first 2-3 digits are the country header, the next 4-6 digits the company prefix, followed by 3-6 figits representing a location, and finally a check digit at the end. The exact numbering structure depends on the GS1 organization that issues the number. Using the GLN, look-up systems at each manufacturer or retailer match the code to a physical location.

The system used to express the location associated with the GLN could be in a number of different formats depending on the preferences of the firm. One standard that might be helpful is GLN AI 414, which identifies shelf or rack locations within a warehouse. Manufacturers and retailers use this standard in conjunction with forklift-mounted readers that scan bar codes placed at rack or shelf locations during put-away or removal of merchandise. This detailed location information is then transmitted to warehouse management systems, providing real-time updates for location and inventory.

The use of bar code standards to add a location component to the EPC read event might be the only option available in the near term. However, the nesting of existing standards into the EPC read event raises a number of longer-term questions concerning overall system design. If several different pre-existing standards for location become part of the EPC read event, the opportunity for interoperability diminishes rapidly as additional translation must take place between different location systems. Building point-to-point translations is time consuming and prone to error. As the volume of EPC transactions increases, it might become impossible to make all of the translations.

Recognizing the importance of location, and its impact on EPC data interoperability, EPCglobal has initiated the Data Exchange (DE) Work Group to address the issue. The final section of this chapter gives a brief overview of the group's preliminary findings

An Approach to Location

When a tag passes within the field of a reader, the serial number of the EPC combines with the location code for the reader, forming a payload of data that also includes a time stamp. This is a read event. The data payload from the read event then travels through the middleware for use in various enterprise systems

The current TDS specification assumes that legacy numbering systems will be embedded within the data payload, either as part of the EPC or the reader code (location and time). This means that in addition to the reader location, the payload will also have the ability to carry extra information about shelf location through the GLN.

For many industries, having a second identifier for location as part of the payload offers another incentive to become a member of EPCglobal Network. These industries include automotive, healthcare, aviation, and defense, which all have existing numbering schemes for location that involve shelving systems.

The use of the GLN in combination with the reader location code also allows for the possibility that the reader might be in motion (affixed to a forklift). In this case, the reader code does not correspond to a fixed position unless a Global Positioning System (GPS) provides additional real time data. The GLN could be associated with the GPS data, providing an accurate location for a forklift within a warehouse. Alternatively, the GLN might map to a fixed location, allowing a mobile reader to accurately sense a RFID tag affixed to a shelf or warehouse racking system.

Though the potential uses for a second location code within the EPC payload data are still under consideration, the possibilities of using specific location information are fundamental to the advancement of automation in warehouses and the ability to send instructions to a specific object located within a supply chain.

Achieving this level of interoperability and automation will only be possible through the detailed work of standards bodies such as EPCglobal, fundamental research on the issues, and objective input from industry practitioners.

PART II: LEVERAGING THE SUPPLY CHAIN: CASE STUDIES

CHAPTER 5

Warehousing: Improving Customer Service

The reality of modern business is that "all competitive advantage is tempo-rary."[1] To remain viable, firms must continually focus on current capabili-ties along with building new competencies for the future.[2] This process of renewal is a multi-dimensional activity that requires diligence in pursuing opportunities where cost savings are apparent, such as improvements in manufacturing efficiency or reduction in the cost of raw materials. How-ever, other areas within business also offer significant opportunities for competitive advantage based on subtleties that are no less important than cost savings.

One such area is warehousing and customer service. Typically considered the primary interface between a firm and its customers, warehousing and customer service represents both a significant cost center and an important element of the marketing mix. These activities require precise execution to achieve a required level of competency that customers expect.

Within supply chains, warehousing plays an important role in the storage of finished goods.[3] All physical objects are stored in warehouses at least one time prior to sale. The storing, handling, and shipping of items involves many manual processes that contribute to cost and the ability to provide service.

Though it is a separate discipline within supply chain management, cus-tomer service relates to warehousing through the goal of ensuring the effi-cient hand-off of goods between a firm and its customers. Often focusing on numerous details, the delivery of acceptable customer service depends on knowledge of products and business processes,[4] proper systems to pro-vide relevant information, and a well-conceived supply chain design.

Warehousing and customer service provide the operational parameters for a good relationship with the customer. In the case of consumer goods (CG) manufacturing firms, the customer is the retail trade. Many large retailers such as Wal-Mart place significant emphasis on the efficient transfer of goods from vendors and cost-effective storage and handling within their own facilities.[5]

The EPCglobal Network and RFID technology offer a number of opportunities to improve warehousing and customer service through new types of information and methodologies. During the past 30 years, identification technologies such as the bar code have made a significant impact in reducing warehouse cost and improving throughput. However, as one researcher has noted, "...the efficiency gains from the use of bar codes have largely been achieved and now the industry is looking for the next generation of AIDC [automatic identification and data capture]."[6] Future advances in warehousing productivity will probably come from the elimination of the "human element in gathering data," an area in which RFID can make significant contributions.[7]

Taking an initial step to analyze the opportunities of RFID requires a fundamental understanding of the customer interface. The next section explores some of these basics.

The Value of Customer Service

While customer satisfaction is generally defined as the output of the total marketing effort, customer service is the direct output of supply chain activities such as ontime delivery to the customer and order accuracy.[8] An essential part of the marketing effort is satisfying the customer while earning a profit.[9] The basic business model of the consumer goods and other industries is that a satisfied customer will make repeat purchases, thus creating long-term economic value through building a recognized brand.[10]

In building brand awareness, four marketing activities create customer satisfaction. These are product, price, promotion, and place. The first three are classic marketing functions, however, the fourth, place, is associated with customer service and supply chain management.

Evidence exists that the four "ps" do not contribute equally to market share. In at least one study, product and place contributed more to increases in market share than price and promotion for a specific situation involving furniture distributors.[11] These results have been validated by additional empirical studies, which found among other things that customer service has a direct link to market share.[12]

Given the importance of customer service, concentrating marketing efforts solely on product, pricing, and promotion would be a disastrous strategy for firms to follow.[13] Because of an awareness of the value of customer service in the consumer goods industry, many firms have built significant competencies in this area. However, like marketing activities, it is difficult to establish the exact contribution of customer service to improved sales. While it appears to be true that comparable levels of customer service relative to a competitor do not seem to increase sales, it does seem to be the case that poor customer service will inhibit sales.[14]

With customer service being an important part of the marketing mix, it makes sense to examine the prospects of applying RFID to create improvement through better information flow. The first step in uncovering these opportunities involves an understanding of the specific businesses processes that make up customer service. Although each business and industry is different, there are some commonalities. An early research project identified the elements of customer service considered important for a specific industry.[15] The following is a list of these elements:

- Ability of manufacturer to meet promised delivery date
- accuracy in filling orders
- advanced notice on shipping delays
- order cycle consistency
- length of promised lead time for quick ship orders
- accuracy of manufacturer in forecasting estimated ship dates
- completeness of contracted orders
- ability to expedite and/or provide rush service
- completeness of quick ship orders
- action on customer service complaints
- manufacturer's willingness to accept returns of damaged product
- manufacturer pays freight and handling returns

A common theme for all of these elements involves speed, accuracy, reliability, and the ability to get detailed information about specific orders (and products) quickly. These are all areas where RFID implemented within warehouses along with unique identification can make a significant contribution to the improvement of customer service.

Because customer service deals with some subtle aspects of the business relationship between a vendor and a customer, documented outcomes of poor service and the resulting effect on revenues and profits are sometimes hard to find. However, one example presents an interesting case.

Considered a fundamental aspect of customer service, the billing process is a typical instance where details can have a big impact. In one situation, a major corporation redesigned its entire billing system to organize around three business lines. The transition caused a great deal of confusion and a large increase in the number of billing errors. By one account, "salespeople were spending as much as 80% of their time sorting out billing problems" drastically reducing their ability to seek new business.[16] In addition to customers being upset, many refused to pay. All of these developments served to cause a severe crisis for the company and sales decreased a substantial amount. Acknowledging that poor customer service was the root issue, the company took radical measures to reverse the damage.

Even though customer service often carries a low profile, it can cause significant amounts of business disruption if the proper information does not exist at the right time to satisfy customer needs. Since many customer service elements link directly to warehousing operations, it is important to look at basic activities, such as billing, receiving, put-away, picking, and transferring for improvements that utilize RFID and the EPCglobal Netwok.[17] "By automating manual, error-prone tasks common in the consumer goods industry, companies can better manage their supply chains" and provide improved customer service.[18]

To date, much of the focus on achieving greater efficiency and accuracy for warehousing and customer service has focused on establishing and monitoring process metrics, continuous improvement programs, and benchmarking with other companies.[19] While these activities will achieve results in practice, there are future technological alternatives involving RFID that have the potential to surpass the strides made by bar code implementations within warehousing.

Automating Warehouse Operations

Assuming that passive RFID technology will become economical for tagging of cases and pallets, there are a number of warehousing tasks that can be automated to speed the flow of inventory and provide better service to customers.

For example, it is common that workers reconcile shipments against bills-of-lading, packing, and pick lists by using physical counts and manual data entry. Having humans perform these tasks causes friction at transition points within the supply chain leading to errors and slower processing times.[20] By removing friction points using RFID technology, firms will better serve their customers with more accurate orders. As noted, delivering superior service creates customer loyalty and eventually contributes to increased profit and market share. For the future, RFID technology holds great potential in reducing supply chain friction at all levels.

The need for accurate, real time data about products in the supply chain is particularly important in the CG (consumer goods) industry. During the past 20 years, CG firms have made significant gains in improving manufacturing methods. This has reduced the cost of many products. CG companies are now looking to their supply chains for future cost reductions. The CG industry moves large volumes of many different products through complex supply chains that often span great distances. Despite the success and widespread use of bar codes, it is becoming clear that the CG industry needs more specific information than what bar codes can provide to manage the flow of goods. Additional information might include:

1. Complete bill-of-lading information attached to each case or pallet.
2. Manufacturing details, lot, plant and quality information
3. Technical aspects of the product, including proper usage.

All of these are essential parts of customer service. To demonstrate the potential of using RFID in warehousing and customer service, this chapter concludes with an illustration drawn from the CG industry. The example was part of a MIT Master of Engineering in Logistics thesis project conducted on-site at a warehouse located in the Boston area.[21] The firm is well known in the CG industry, and will be referred to as "Company A" to maintain confidentiality.

Efficient Handoffs in Order Fulfillment

By looking at typical order delivery processes at Company A, we can explore how RFID will affect warehousing. In shipping customer orders, goods are constantly transferred from one stopping point to the next. For example moving inventory:

- From a distribution center onto a truck.
- From the truck into a trucking terminal.
- From the trucking terminal onto another vehicle for transport to the retailer or retailer's distribution center.

At each transfer point, personnel perform a series of counts and documentation tasks. These tasks have two components that affect warehousing. The first is a labor cost, because humans are needed to perform these tasks. This introduces the risk of error and exposure to loss through theft. The second disadvantage of manual processes is the friction it causes in slowing down the flow of goods. Each time a worker must manually check and record data to process an order impedes the movement of goods.

Using RFID Technology, Company A has implemented a prototype portal approach (readers located on the loading dock) where the goods pass near a scanner to verify the type and quantity during loading into a truck or railcar. The results are compared against the packing list generated from a purchase order. If an exact match does not occur, the system can automatically alert the distribution center staff at Company A about the discrepancy. This RFID enabled process eliminates the need for manual counting and improves customer service by eliminating shipping errors.

In addition, since Auto-ID technology does not rely on a direct line of sight to capture information, as is the case for bar codes, the labor involved in rearranging and orienting goods for scanning is eliminated.

By decreasing the number of times inventory or shipments are manually verified, Company A can increase supply chain velocity. Looking at a specific situation, the time spent checking goods and reading bar codes, was over 11,000 hours per year. RFID would eliminate the need for these manual procedures.[22]

The process of counting, checking, and reconciling the goods does not stop at Company A's distribution center. These tasks are repeated many times before goods arrive on retail shelves. For less than truckload shipments, goods are typically sent to a consolidation hub. Here the goods are removed from one truck and staged with other items. The new combination of goods is then loaded onto another truck for transport to either a retailer or a retailer's distribution center. Each time the goods are moved at the terminal, the trucking company performs a count for confirmation of the bill of lading. There is also a packing slip and freight record for each order placed on a truck that must be reconciled.

During the consolidation process there is constant risk that the goods will be left behind or loaded onto the wrong truck. This results in lost inventory and poor customer service to retail outlets through late and incorrect deliveries. There are many opportunities to apply RFID for all of these basic processes.

A Deeper Look

By installing RFID systems, Company A can quickly count goods when they are ready for shipment, eliminating the need for manual counts. Manual counting and verification is purely an error reduction process that adds no value. With RFID, companies have the potential to quickly and accurately trace an item anywhere in the supply chain as long as there are readers to detect its presence.

For example, several readers could be situated on loading docks near truck bays. As forklifts move goods into trucks, the readers' could automatically scan and record the amount loaded. Orders can be instantly checked to ensure that the shipment has the right products and the right quantity. RFID reveals the problem of shipping the wrong unit or incorrect quantity of goods at the point of origin rather than at the end. By having the confirmation at each step, errors in shipments are caught along the way, when they are cheaper to remedy.

RFID simplifies the shipping process and reduces the supporting documentation required to fulfill an order. This decreases the chance that a document will be misplaced or forgotten altogether. The instant verification along each major transition point can locate where an order was compromised. Manufacturers thus gain the advantage they need to bill correctly and reconcile disputes at each point in the supply chain. By automating these tasks, companies reduce friction in the supply chain. The goods flow faster, thus reducing the order cycle time and allowing for faster inventory turns and better customer service.

Other Areas for Improvement

In addition to the order fulfillment process and associated shipping operations, there are other areas where RFID can have a potentially positive
impact on warehousing and customer service. For Company A these areas
include reduced invoice deductions, automated inventory cycle counts,
and enhancements to the returns process. Table 5-1 shows the projected
dollar savings for a single distribution center operated by Company A.[23]

Table 5-1 – Cost Savings in Warehousing

	Hard Savings	Additional Benefits
Efficient Handoffs	$1.0 MM in labor costs	reduced processing times
Reduction of Invoice Deductions	$2.5 MM in write downs and capital cost	better customer service
Automated Inventory Counts	$0.1 MM in labor costs	reduced year end write downs
Efficient Returns Management	$1.0 MM in labor costs	faster returns processing
Total	$4.6 MM	

Though the study of Company A did not include a detailed assessment of
RFID deployment costs, the fact that more than $4 million in savings could
result from streamlining activities using RFID attracted the attention of senior management. This savings represented 0.2% of the sales throughput for
the distribution center, a significant amount for the CG industry where
profit margins are thin.

Of particular interest to management was the savings associated with
deductions that resulted from inaccurate counts and invoicing errors. Better information and the elimination of manual counting, both capabilities of
RFID, could reduce a good portion of this cost.

Besides the cost savings, RFID has potential to create value that often does
not show up on the balance sheet or income statement. Sometimes called
"organizational assets," capabilities such as short lead times and speed of
delivery are intangible yet these activities have significant financial value
for firms such as Wal-Mart and Dell.[24] Creating new forms of organization
assets might be the greatest value of RFID in supply chain management.

As a concluding note, supply chain friction caused by small, time-consum-
ing activities that slow the velocity of goods, has national implications in
terms of cost. By one estimate, the security measures implemented after the

9/11 attacks on the United States resulted in an additional $151 billion in cost because a number of verification activities slowed the velocity of goods and services.[25] As RFID expands in use, it will become a powerful force to uncover and reduce these hidden costs.

CHAPTER 6
Maintenance:
Service Parts Inventory Management

Perhaps one of the most promising areas for application of the EPCglobal Network and RFID technology involves service parts inventory management. Often associated with high value items such as critical components for aircraft or computers, service parts inventory management plays an important role in providing maintenance support once various machines leave the manufacturing facility and become operational in the field.

Though manufacturers have focused a great deal of emphasis on increasing reliability, there continues to be a need to stock service parts in the event of a product failure. To limit machine downtime, many suppliers have a policy that guarantees extremely high service levels, making parts available almost all of the time. For example, leading companies like Caterpillar ship 99.7% of parts ordered within one day to virtually any area of the world where their equipment is in use.[1]

Since service parts are often expensive, it becomes important to make the proper tradeoff between inventory investment and service levels to minimize cost. One of the main factors in calculating the amount of service parts to keep on hand is an accurate forecast of component failure rates.[2] Going the extra step of gathering information to improve the forecast is especially important under the following conditions:[3]

– Short product lifetimes
– High sales levels
– Products sold during longer spans of time
– Seasonal product sales
– Desire for high customer service levels
– The cost of a shortage is high and the cost of holding is low
– The cost of collecting the information is low

From a practical standpoint, the information needed to do an accurate forecast is often unavailable. The EPCglobal Network along with RFID technology holds the promise of new ways to generate information for improving service parts forecasting. This is especially the case when there is a known installed base of machinery and formal maintenance agreements exist to provide service parts when machine failures occur.

With a focus on a real-world situation, this chapter examines the opportunity for implementing the EPCglobal Network and RFID technology in a specific service parts business. It is through the examination of business processes that practitioners can gain a clearer picture of the benefits and challenges associated with the EPCglobal Network and RFID technology.

Overview

The case study outlined below involves service parts logistics operations for a company that supports an installed base of computing machines.[4] The company offers hardware and software support for network servers and workstations housed on both UNIX and Windows platforms. Customers predominantly are Fortune 500 firms with massive information processing infrastructures. To provide quick service when component failures occur, the company maintains a network of parts banks in the United States and Europe. Coordination of this network is done at a central call center that is the initial point of contact for all customer service requests. The company also employs field engineers, both directly and under contract, in each of the geographical areas that it provides service.

Mid-sized and headquartered in the Boston area, primary competition for the company comes from the service parts operations of large OEM vendors in the computer industry. Privately held, the current owner was also the founder.

Since its inception, the company has maintained a competitive strategy based on three policies: 1) high service levels, 2) flexible service plans to meet the needs of a wide range of customers, and 3) a single point of service for equipment manufactured by different vendors. The commitment to deliver flexible service plans differentiates the company from OEM vendors that tend to offer a standardized service plan for all customers. This strategy has led to impressive revenue growth during the past decade.

However, with the economic downturn beginning in 2001 the company has experienced declining profit margins primarily from the rising cost of operations. This happened because of an increase in the complexity of the installed base of network servers and workstations. In addition, increased competition from OEM vendors forced downward pricing pressure for new service contracts further squeezing profit margins. Rising operational costs, invigorated competition, and the lack of pricing power in the market led the company to conduct a review of current operations and established practices.

Business Operations for Service Parts

The following subsections discuss the fundamental aspects of the business including the structure of service contracts, the evaluation of new contracts, and the typical process flow for a service request from a customer. The section concludes with a discussion of the main issues that caused the recent decrease in profitability.

Contracts

Detailed contracts are negotiated for all business conducted by the company. A typical service contract specifies the following:

- *Types of equipment supported* – This includes a listing of the make and model of the computing machines supported at each location. There is usually no indenture level information for any of the equipment. Much of the information captured in the contracts is incomplete. Updates are infrequent.
- *Term of contract* – Length of time the contract is in force, usually one or two years.
- *Type of service* – hardware or software support, or both.
- *Guaranteed response times* – The contract specifies the maximum allowable response time for a service request at each location. The most frequent choices for response times are same day (2, 4, 6, and 8 hour response) and next day.
- *Penalty clauses* – Not meeting guaranteed response times for a pre-specified percentage of the requests results in a penalty. This pre-specified percentage is the contract service level. Usually, the service level is between 80% and 90% and is specified in the contract.

Before entering into each service contract, the company performs a detailed evaluation of the customer's requirements, taking into consideration the existing capabilities of the parts bank network. This evaluation results in an estimate of the incremental profit contribution for each potential new customer. A flow diagram of the evaluation process appears in Figure 6-1.

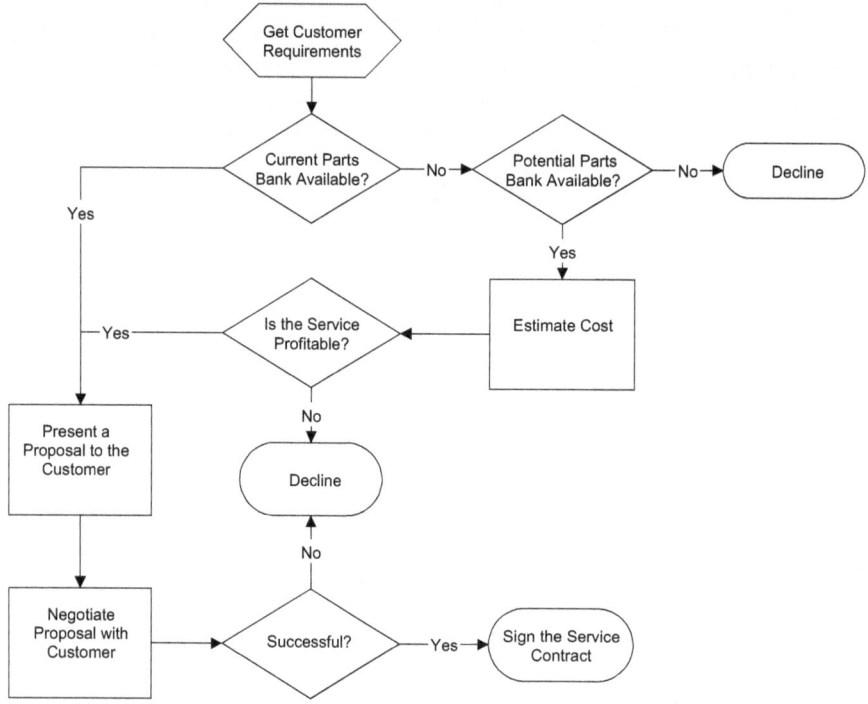

Figure 6-1 – Business Process for Evaluation and Signing of Service and Support Contracts with Customers

Because the company has done a number of service contracts in the past with profitable outcomes, the process described in Figure 6-1 is an established practice. Though this process provides a structure to ensure profitability during the life of the contract, it depends upon information about the installed base. Often, during the lifetime of the contract, the installed base changes causing additional complexity, lower service, and increased costs. Once the contract is in place, it is seldom renegotiated because of a change in the installed base. This presents a significant management challenge for the company to overcome.

Service Network

The service parts network is a two-echelon distribution system consisting of a central depot and 50 regional parts banks maintained by third party logistics providers (3PLs). These facilities serve about 2,000 customer sites. The 3PL providers make local delivery of parts to customer sites. The company also maintains parts banks onsite at the customer location for contracts that guarantee a two-hour response time. A min/max (s, S) continuous review replenishment policy is the method used by the company to determine reordering for all of the parts banks. All defective parts replaced at customer sites are shipped back to the central depot for repair and testing before being returned to stock.

Since most of the service parts have similar weight, the transportation cost depends on the distance between the parts banks and the customers. Parts banks are usually located within 200 miles of the customer site if the customer requires one-day part delivery service.

Inventory Planning

The safety stock levels at the parts banks are determined based on the historical demand rate. This simple approach also serves as the forecast. For lower demand items, the safety stock level is simply the demand during the previous month. For higher demand, the safety level is two times the demand during the last 15 days. These rules apply to both the parts bank and the central depot. For a new part where no historical demand data exists, the company uses the safety stock level recommended by the manufacturer of the part.

Fulfillment Process

Once a service contract is in force, the company maintains a disciplined fulfillment process described by Figure 6-2.

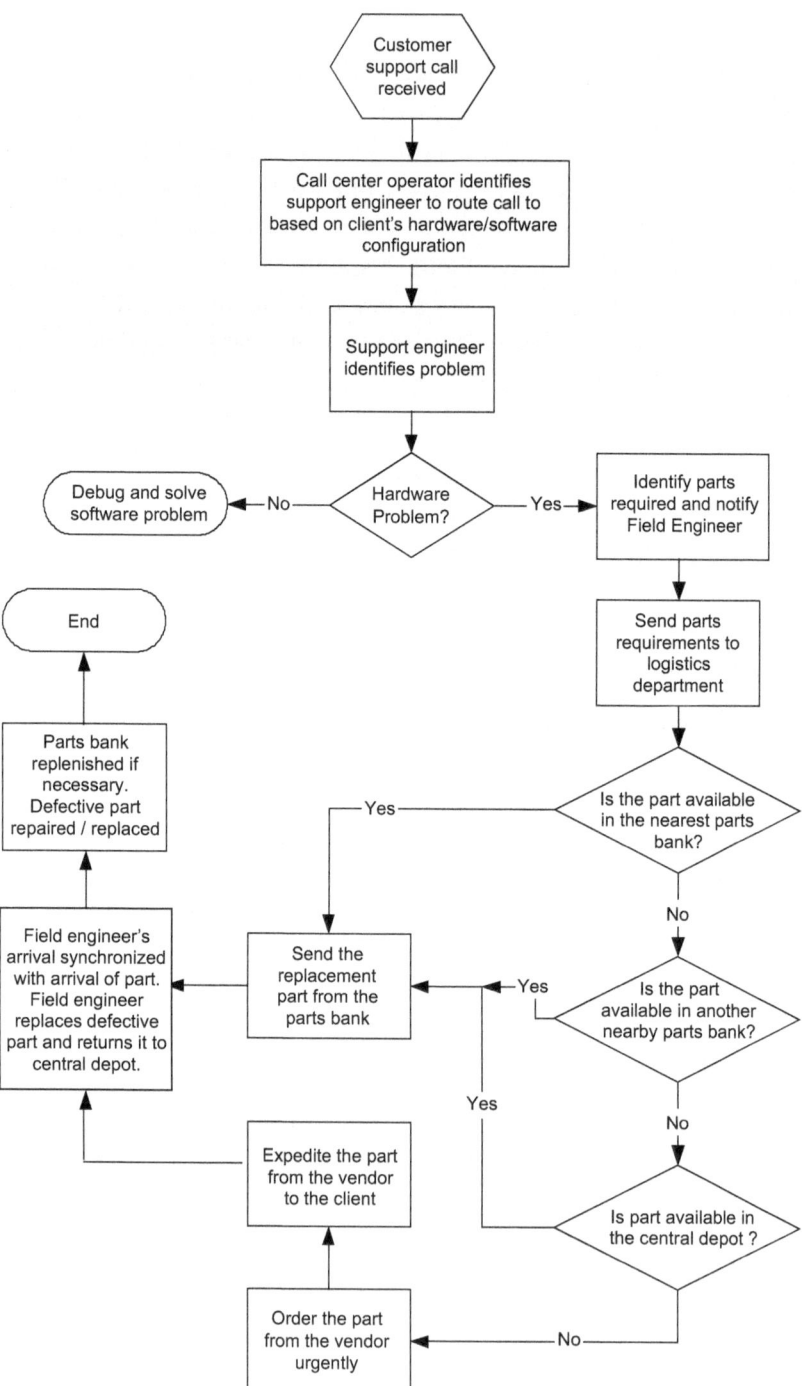

Figure 6-2 – The Service Operation

This fulfillment process is the critical aspect of day-to-day operations. To meet the specified service levels, the right part must be at the right place at the right time. As a result, a significant investment in inventory exists. Managing this critical asset is the most important variable in maintaining profit margins. The company has found that it is easy to over invest in inventory as a quick means of increasing service levels. With increasing complexity of the installed base, this leads to reduced cash flow.

Critical Business Issues

Like any business, the company faced a number of important operational issues. Three primary issues were thought to be significantly affecting the profitability of the company.

- **High Inventory Cost, Low Service Level** – Inventory carrying cost for all the parts located in the Parts banks and the Central Depot was 25% of annual service contract revenue. Average parts availability was less than 80%. This compares unfavorably to other companies in the computer service and support industry based on survey results (see Table 6-1).[5]

Table 6-1 – Performance Metrics

COMPANY	Inventory Investment/Service Revenue (%)	Service Level (% On-time Parts Availability)
1	10%	92%
2	23%	88%
3	40%	88%
4	17%	90%

- **High Scrap Rate** – Because of obsolescence, the company scrapped approximately 4% of the service parts inventory cost per month.
- **High transportation costs** – The annual transportation cost equaled 15% of the total value of the inventory. In contrast, a survey finds that the average transportation costs for companies ranged between 2.2% to 11.4%.[6] Stock outs on frequently used parts resulted in high expediting costs. In addition, there were significant lateral transshipments between parts banks to meet backorders and correct stock allocation based on recent demand.

These three issues produced downward pressure on operating margin. With decrease profits, the company was willing to consider new methods that could improve productivity.

Suggested Solutions (Non-RFID)

To address the critical business issues, the firm needed to make better trade-offs between inventory carrying cost and service. Given the complexity of the service parts network, these cost and service trade-offs required implementation of mathematical models for effective solutions.

Most would accept that traditional time-series methods for calculating forecast demand, and in turn optimal inventory levels, do not work for service parts because the demand is "sparse." This means that demand is often zero for many periods. In this case, specialized models are needed to predict service parts demand.

From the initial review of operations, a recommendation to implement two models surfaced: METRIC developed by the RAND Corporation and used by the Air Force, and a location model to optimize the number and location of parts banks [7, 8, 9; 10] Both of these models are established methods that optimize cost and service trade-offs. Through implementation of these models, the company would realize significant cost savings.

However, METRIC and the location model did not address the issue of a lack of information about the installed base. A full understanding of the installed base covered under the service contract is critical in determining projected demand. The EPCglobal Network and RFID technology provide a powerful means to gather this information.

A Deeper Look at Service Parts Demand

At the most basic level, demand for service parts arises from the failure of one or more components in a computing machine. For purposes of managing inventory, the forecast should consider only failures that occur during the useful life of the part. Warranties usually cover initial malfunctions during the "burn-in" phase. Because of rapid obsolescence for computing machines, wear-out is hardly an issue. Few components reach the end of the operating life prior to the replacement of the computing machine.

Excluding burn-in malfunction and wear-out, the remaining failures that occur during the useful life of a part happen because of "chance" occurrence. In this case, sudden breakdowns take place, sometimes without advanced signs of pending failure. Machine load does have correlation to component failures in some cases.

Hence, the overall demand rates for service parts depend on 1) the number of machines being supported (the *installed base*) and 2) how much these machines are used (the *program usage*). However, service providers for computing machines, like the company in this case study, often have only a general idea of the installed base, and program usage.

Because of this lack of information, there is no alternative but to rely upon time series data of failures as a means of calculating safety stock levels. The problem with this method is that there is often a significant time lag between a change in the installed base, or program usage, and a change in the failure rate. In addition, it is very difficult to correlate the installed base, measured in machines, to failures measured in demand for parts, because some components are interchangeable with other machines at the same location. Adding to complexity, service support organizations typically stock thousands of parts making manual review of changing failure rates difficult to accomplish.

The EPCglobal Network and RFID technology offers a solution through the prospect of identification and monitoring of the entire installed base. The next section discusses some of the aspects of RFID technology that are important to increasing productivity in the service parts industry.

RFID Enhancements

Though there are few current applications of RFID technology in service parts inventory management, several opportunities exist to apply the technology. In the context of the case study presented in this chapter, two basic applications make sense.

1. *Monitoring* – Perhaps the most important aspect of RFID related to service parts inventory management is the capability of monitoring the installed base and program usage within computer facilities. Given the lack of information in these areas along with the high value of service parts needed to support complex computing systems, monitoring will almost

certainly reduce the amount of service parts needed to achieve target service levels.

Assuming that critical components contain RFID tags, either applied externally or integrated into the electronics, there are two basic ways of monitoring the installed base, intermittent or continuous. First, a technician could manually scan machines with an RFID reader to gain an accurate understanding of the installed base of components. This is essential information for projecting the number of probable failures for a span of time, thus improving the forecast. With a better forecast, the proper amount of inventory to meet a specific service level can be reserved.

A second way of monitoring involves a continuous scan of the installed base using readers that are permanently fixed in the customer's facility. Besides confirmation of the installed base, sensors could be added that would monitor such variables as electric current draw to determine the time of usage for a component and confirm it is operational. This approach is similar to that discussed in Chapter 10 (food) for sensing technology. In addition, the consumer goods industry already uses "in-store" monitoring, a topic discussed in more detail as part of Chapter 15.

Both methods, intermittent or continuous monitoring, allow the service parts supplier a means of linking directly to the customer with the longer-term prospect of gaining real-time information about critical components. The EPCglobal Network and its ability to organize serial numbers to achieve unique identification will serve an important role in enabling this link so that specific information can be gathered for an individual component.

2. *Repair Operations* – Beyond ability to gather information about the installed base and program usage, the EPCglobal Network and RFID technology also has potential to redefine repair operations in terms of productivity. Often customers return high value, non-functional components to a central facility for repair. After the repair work is completed, these components then re-enter the inventory of service parts for future use.

Attaching a RFID tag to each component before shipping to a central repair facility would greatly improve the process of tracking. There are anecdotal accounts where considerable productivity can be achieved from knowing the precise location of high value components.

The telecommunications industry offers an example.[11] Network cards used in telephone switching operations often cost $35,000 or more, yet have small dimensions and can be easily lost. Occasionally, these network cards fail and are returned to a central repair facility that occupies 10,000 square feet. Though central facility workers scan a bar code placed on each network card upon receipt, there are situations where repairs take place and the location of individual network cards becomes unknown. This happens because the bar code only identifies the product type and not the individual item. Since each item requires different types of repair work, it becomes important to have unique identity.

Further, even though the facility had bar coded racks, and location of the network cards was known, the specific part at a specific location was unknown. By one account, workers often spent up to a day trying to find a specific network card within the repair facility.

By affixing an RFID tag at the time of shipping, unique identity for a component can be maintained throughout the repair process. This has a number of advantages in tracking the exact location within a facility and the amount of repair labor expended per individual component.

While the above recommendations for using RFID technology have not yet been implemented by the company described in the case study, there are situations where firms have aggressive schedules to implement RFID for service parts on a large scale.

In 2005, "Boeing announced it would require many of its suppliers to begin placing RFID tags on a number of parts used in its latest line of commercial airliners, the Dreamliner 787." [12] During the announcement, Boeing mentioned that the company believed tagging would make it much easier to track the maintenance histories. As planned, the tags would remain attached for the entire life cycle of the part.

A further extension is to use RFID for unique identification at the end of the life cycle for a product [13] Unique identification would also be helpful in recycling and handling of hazardous materials.

CHAPTER 7

Pharmaceuticals: Preventing Counterfeits

The complexity of the United States health care system is increasing rapidly. Demographic changes, along with a host of new drugs, are causing greater volumes of raw materials and finished products to move through the pharmaceutical supply chain. In many ways, the pharmaceutical supply chain is beginning to resemble the distribution of consumer goods.[1] However, several important differences remain.

The fundamental goal of the medical industry is patient care and safety. To achieve these goals for the public good, the Food and Drug Administration (FDA) and individual States regulate the industry through laws and administrative orders designed to protect the integrity of drugs throughout the pharmaceutical supply chain. Specifically, the Prescription Drug Marketing Act (PDMA) of 1988 calls for transactional history to follow individual packages of drugs that travel from manufacturers through wholesalers and finally to retail outlets such as pharmacies and hospitals.

The FDA uses the term "pedigree" to denote the requirement for documentation of transactional history. The pedigree requirement, legislated as part of the PDMA has been on hold because of practical limits in terms of implementation. Determining a pedigree for all the drugs in circulation would require millions of pages of information to document the flow of drugs from manufacture to consumption. [2]

With a recent renewed threat of counterfeit medications entering the US pharmaceutical supply chain, greater interest exists in achieving full implementation of pedigree laws first outlined in PDMA. While the number of counterfeit cases in the US remains below that of developing countries such as India and China, an increase has occurred from an average of five cases per year prior to 2000, to over 20 cases per year currently.[3] By some estimates, counterfeit is as much as 80% of all drugs sold in some third world countries." [4]

Since the US pharmaceutical market is vulnerable to fake drugs, manufacturers have introduced a variety of measures in an attempt to thwart counterfeiters. All of these technologies share the common strategy of placing some indicator on the package, or the drug itself, as a means of verifying authenticity.

However, it appears by using various computer and packaging technologies, counterfeiters have the ability to quickly imitate these measures and infiltrate counterfeit drugs into the legitimate drug distribution chain. By one account, counterfeit and theft combined cost the pharmaceutical industry $30 billion per year.[5]

Implicit in the documentation process outlined by the PDMA is the administrative requirement to do track and trace. *Tracking* involves knowing the real-time physical location of a particular drug and its future path within the supply chain. *Tracing* is the ability to know the historical locations, the time spent at each location, record of ownership, packaging configurations, and environmental storage conditions for a particular drug.

Track and trace forms the foundation for improved patient safety by giving manufacturers, distributors, and pharmacies a systematic method to detect and control counterfeiting, drug diversions, and mishandling. These are important aspects of supply chain security. Unfortunately, the current system for the documentation and organization of data is cumbersome because of a reliance on manual procedures and storage of information on paper. As a practical result, track and trace takes place only in an emergency such as a drug recall.

The EPCglobal Network and RFID technology offers the prospect for an integrated solution to the track and trace problem, providing a means of reducing the amount of counterfeit drugs within the United States pharmaceutical supply chain. The open standards feature of the technology aids in the implementation of a supply chain wide application. In addition, the EPCglobal Network sets the foundation for a number of other applications within the health care industry that include continuous patient monitoring, shared yet secure medical records, valid and accurate medical dosages, medical equipment tracking and improved information display and communication.[6]

The next section examines the scope of the counterfeit problem and the legal underpinnings for improved trace and trace capabilities within the pharmaceutical supply chain.

An International Problem of Significant Magnitude

The World Health Organization (WHO) defines counterfeit as: "A medicine that is deliberately and fraudulently mislabeled with respect to identity and/or source. Counterfeiting can apply to both branded and generic products and counterfeit products may include products with the correct ingredients or with the wrong ingredients, without active ingredients, with insufficient active ingredients or with fake packaging." [7]

According to the WHO definition, what makes a drug/medicine counterfeit is the deliberate or intentional (criminal) nature of the mislabeling or adulteration of a drug. This type of illegal behavior leads to 1) compromises of patient safety, 2) economic loss to established drug manufacturers, and 3) a threat to the national security of sovereign countries.

The WHO estimates that between five and eight percent of the worldwide trade in pharmaceuticals is counterfeit.[8] Many industry experts believe this to be a conservative estimate. Anecdotal reports indicate a significant increase in counterfeit drugs during the past few years.

A few examples of published articles on counterfeit include:

Up to 33% of anti-malarial drugs for sale in Cambodia, Laos, Burma, Thailand and Vietnam contained no active ingredient." [9]

"Approximately 192,000 people died in China in 2001 due to the effects of counterfeit drugs. As much as 40% of drugs in China are counterfeit." [10]

"In Columbia, up to 40% of medications are believed to be counterfeit."[11]

"Approximately 50% of drugs sold in Nigeria are counterfeit." [12]

In a study conducted simultaneously at Dulles and Oakland International Airports, U.S. Customs and FDA agents found that 10% of the drugs they analyzed contained no active ingredients.[13]

The problem of counterfeit drugs has reached grass roots America. Pharmacist Lowell Anderson of Bel-Aire Pharmacy in White Bear Lake, MN states, "I have been in this business for 40 years...I have less confidence in the integrity of the supply chain today than ever before. It scares me."[14] This appears to be a well-founded concern. During the past ten years, drugs sold in America such as Procrit, Epogen, Serostim, Zyprexa, Diflucan, Combivir and Retrovir have been counterfeited. Even Lipitor, the widely prescribed drug to control cholesterol levels, was recalled because of a counterfeiting incident. In this particular case, the FDA could not determine how many bottles were in each of three counterfeit lots. In addition, the counterfeit lots could not be accurately tracked or traced. While most counterfeit drugs contain harmless ingredients such as water or glucose, the counterfeiting of Lipitor "posed a potentially significant health hazard" according to the FDA.[15]

Even though the overwhelming majority of drugs sold in the United States are safe, the $220 billion per year pharmaceutical market is an attractive target for counterfeiters.[16] With the complete mapping of the Human Genome, there will be a number of new, high priced drugs appearing on the market during the next few years. This will increase the opportunity for counterfeit.

The Causes of Counterfeit

Three factors account for the increase in counterfeit drugs:[17] First, the computer technology available to forge labels has become more sophisticated. It is now possible to reproduce any label. Second, there is an abundance of small wholesalers buying and selling medications. Along with differences in pricing, the increase in small wholesalers creates an active secondary, or gray market. In some situations, drugs change hands many times before reaching pharmacies. This increases the opportunity to introduce counterfeit into the supply chain. Finally, an increased number of expensive drug therapies provide lucrative potential for forgers to net large profits. In some cases, organized crime and former illicit drug dealers have entered the counterfeit ethical drug market because the profit potential is so large.

Federal efforts to deal with counterfeit are hindered by laws that do not assign supply chain wide accountability to any one authority. Re-importation and diversion, in addition to the advent of internet pharmacies, makes counterfeit hard to prevent. In summary, it is not very difficult to produce a counterfeit drug for introduction into the United States pharmaceutical supply chain. The incentive is great for criminals to take part in this illegal activity.

The Changing Regulatory Environment

With greater awareness of counterfeit drugs, the FDA and states are moving forward with new legislation to combat the problem. Florida gained national attention when Governor Jeb Bush signed the Prescription Drug Protection Act on June 3, 2003.[18] This establishes the legal requirement for a "pedigree" to accompany each drug sold in the state.[19] The intention of the law is to verify authenticity and reduce the chance of counterfeit. The law calls for a phase-in of pedigree documentation over several years with full implementation for all drugs by July 2006. This introduces a number of important issues for the pharmaceutical industry to consider.

Specifically, the law calls for the following information to accompany each drug through all steps of the supply chain:

1. Drug Name
2. Dosage
3. Container size
4. Number of containers
5. Drug Lot or Control numbers
6. Business Name and Address of ALL parties to each prior transaction, starting with the manufacturer
7. The date of each previous transaction

These requirements add a great deal of complexity for manufacturers and distributors. As an example, the typical drug distributor carries at least 34,000 stock keeping units.[20] Maintaining pedigrees given this volume of drugs is overwhelming with current identification and information technology.

Following Florida's lead, California passed a statute requiring an electronic pedigree for all drugs sold in the state.[21] The California law requires full implementation by January 2007. At that time, all pedigrees transmitted between buyers and sellers must be in electronic form.

Other countries besides the United States have also moved forward with pedigree regulations. Most notably the Italian government, with financial support from the European Union, began in 2000 to enforce the track and trace of pharmaceuticals through the Bollini Law.[22] This law requires application of a special sticker containing a serial number and bar code to each unit of sale.

The Bollini Law also requires all parties within the supply chain to record and archive each serial number. This has created great difficulty for manufacturers and distributors. As a result, the full implementation of the law did not take place until after June 2004 because of a lack of technology to handle the task of recording and archiving the serial numbers. A final design of the database structure needed to accomplish drug tracing had not yet been determined. Open questions remain about the role of the Italian government in controlling and maintaining a large-scale central database containing all drug pedigree information. In an average year, Italian distributors handle 1.2 billion items. It remains uncertain how bar code technology and existing information processing infrastructure will cope with this much data.

Current Solutions to the Pharmaceutical Counterfeit Problem

In response to regulatory and business pressures, the pharmaceutical industry attempts to combat counterfeit using a number of different techniques. To date, no technique has proven effective in totally eliminating the problem. Most detection procedures currently in place rely on manual product inspection by pharmacists or sales representatives to check for evidence of counterfeiting. In the absence of automated inspection technology, these methods are often too costly to do counterfeit inspection on a broad, periodic basis. If positive detection of counterfeit does occur, it is not clear what action to take because current methods provide incomplete information about the scope of counterfeiting for a particular drug.

Recent efforts to deal with counterfeit involve both information and material technologies. Several examples include:

> Some drug companies have injected an inert chemical signature directly into medications, which can be checked with a small handheld device much like a home pregnancy test.[23]

Tamper proof packaging, in addition to technical measures such as holograms, difficult to replicate packaging designs, and unique fonts have been used.

FDA Medwatch is an excellent resource for patient safety information, from label changes to counterfeit product warnings and recalls.[24]

HDMA Product Safety Task Force recommends steps and guidelines the industry should consider for the safe purchase of products.[25]

The Institute for Safe Medication Practices is dedicated to the safe use of medications through improvements in drug distribution, naming, packaging, labeling, and delivery system design.[26]

Product Surety is a joint industry initiative with the FDA to curb the incidence of counterfeiting.[27]

Product anti-counterfeit technologies fall into two broad categories, intra-formulary or package based. There are a large number of these technologies in use today. Table 7-1 lists the most commonly used technologies and an assessment about the chances of defeating each approach. In some cases, certain counterfeit detection measures can be both covert and overt. For example, fibers or threads embedded into a label can be seen with the naked eye or in some cases are invisible without the aid of ultraviolet light.

Table 7-1 – Common Anti-counterfeit Methods

ANTI-COUNTERFEIT MEASURE	COVERT	OVERT	REPLICATION
Intra-formulation			
Imunnoassay	X		Low
Unique Flavoring		X	Low
Package			
Design		X	High
Watermarks	X	X	High
Digital Watermarks	X	X	N/A
Fibres and Threads	X	X	Somewhat - Low
Reative Inks	X	X	Somewhat - Low
Holograms	X	X	High
Bar Code		X	High

Low   – small probability of successful counterfeit
High – increased probability of successful counterfeit
N/A   – not enough information to make a determination

These approaches are static and there are high fixed costs for switching to new anti-counterfeit measures. Manufacturers replicate the approach many times to cover millions of individual dosages for a specific drug. After a particular anti-counterfeiting solution has been in the market for some time, counterfeiters learn ways to defeat the safeguard. Manufacturers constantly have to remain a step ahead by either adopting new measures or mixing and matching current technologies. While the changing of anti-counterfeit approaches helps manufacturers and regulatory agencies differentiate genuine product from false product, the rapid changes are confusing to the public at large. This creates a situation where consumers have a difficult time checking the authenticity of a product themselves.

An Anti-counterfeit System Based on Information

The EPCglobal Network and RFID technology enables two fundamental, supply chain wide approaches to deal with counterfeit drugs. Both of these approaches complement the current techniques employed by the pharmaceutical industry.

First, RFID technology allows the possibility of instant verification for any drug, at any location. This verification process is possible through a proposed information technology infrastructure that spans the complete supply chain. Second, RFID technology provides track and trace capability. This capability supplies true pedigree information about drugs accessible by partners within a supply chain. Together, track and trace, and drug verification, are difficult barriers for potential counterfeiters to overcome.

Both approaches depend on the capability to identify individual drugs within the supply chain at the primary package level. RFID tags, containing an EPC, applied to each unit of dosage provides this capability. To understand the environment in which tags must operate, Figure 7-1 shows a typical pharmaceutical supply chain. Later in the chapter, this supply chain will be used to show how each element of RFID technology, RFID tags, readers, the middleware, ONS and the PML server fit together to form an integrated solution that achieves unique identification of individual drugs.

The RFID approach has advantages as compared to bar codes when doing track and trace or drug verification. Using bar code systems to read the billions of identifiers needed to record location and serial number information for individual drugs suffers from several limitations. First, bar codes

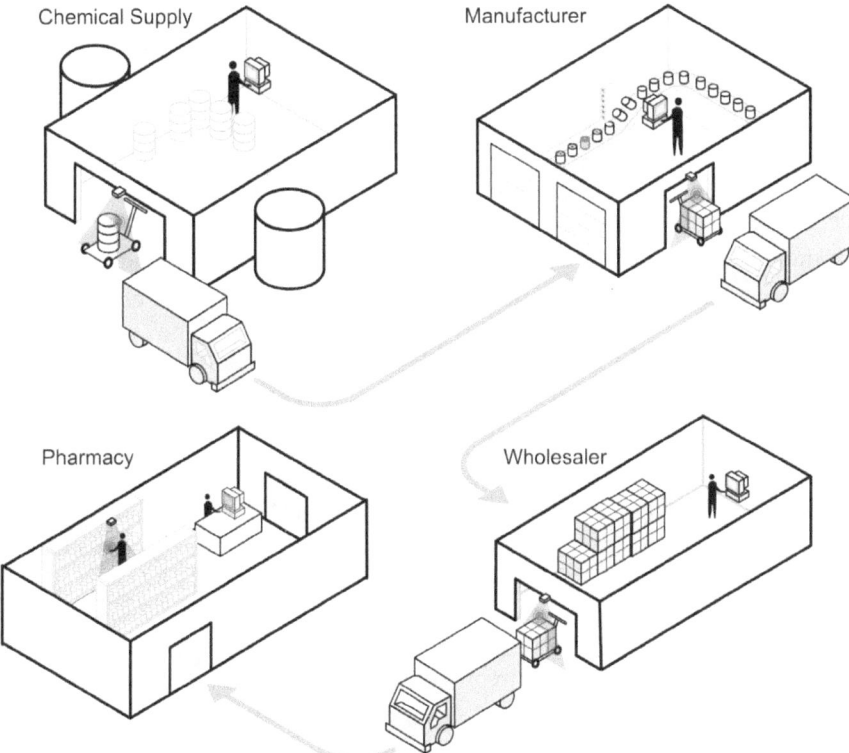

Figure 7-1 – The Pharmaceutical Supply Chain

require a line of sight to do a proper read. For serialized drugs individually labeled with bar codes at the primary package level, and shipped in cartons, direct reads are laborious. In addition, bar codes can only provide unidirectional information, i.e. an item cannot be remotely "asked" to communicate information such as location, or temperature, for recording in an enterprise resource planning (ERP) system database.

The appeal of RFID technology lies in the ability to use the EPC as a pointer to look up important information about a drug that is contained in a remote database. Either the Internet or dedicated computer networks can provide the communication link. This ability to link physical objects to information provides a powerful capability for track and trace, and drug authentication.

However, this new capability does have drawbacks. The task of handling streaming information for billions of individual pharmaceutical products taxes the capacity of the Internet or dedicated computer networks. The next section discusses approaches to overcome this capacity concern.

An Example of Pharmaceutical Supply Chain Complexity

Figure 7-1 shows that the form of the physical goods can change during each step of the pharmaceutical manufacturing and distribution process. Immediately after completion of each step, the product becomes a finished good that continues as an input to the next step in the supply chain.

Referring to Figure 7-1, the finished product for the chemical plant is bulk active ingredient packaged in drums with a specific name, composition, lot number, and expiration date. In contrast, the transport carrier that moves the drums of active ingredient from the chemical plant to the manufacturer sees only a shipment of specific weight and volume. Other attributes are not important to the carrier. There is no direct, continuous link to attributes of the shipment such as lot number or expiration date.

To deal with this situation, pharmaceutical manufacturers have placed select pieces of information directly onto the package by printing bar codes or lot numbers. In this case, the package becomes the vehicle for carrying the information needed for track and trace, and authenticity verification through the supply chain. Though the information carrying capacity of this approach is limited, it does guarantee universal access to all parties within the supply chain.

Unfortunately, this "self contained" approach of physically attaching information to the secondary package can be, and often is counterfeited. In addition, information contained on the secondary package is hard to access quickly because there is no integrated network to organize the data for an entire supply chain consisting of multiple trading partners. This limits possibilities for serialization of drugs at the secondary package level. As a compromise, pharmaceutical manufacturers often rely on assigning lot numbers to large amounts of drugs that might exceed one hour of production. However, this practice lacks the granularity needed for the supply chain of the future.

The identity change for a product continues throughout each step of the supply chain making drug verification, and track and trace, difficult to accomplish even with the self-contained approach for transmitting information. Historically, pharmaceutical manufacturers and distributors have gathered the information needed for drug authentication, and track and trace, using detailed forms and secure databases as storage devices. In even the best situations, this information is difficult to retrieve and seldom shared with other parties outside of the firm. In the event of a recall, special teams within firms are charged with the task of accessing data to make important decisions about the extent of the problem, a labor-intensive process.

Aggregation and Inheritance

Although the physical form of goods changes throughout manufacturing and distribution, a link still exists for all raw materials and work in process used to produce finished goods. This type of link demonstrates inheritance of specific attributes. Each medicine used by the patient has a specific lot number and expiration date printed on the container. The drug is shipped on a specific truck, at a specific temperature for a specific duration. The effectiveness of the medicine ultimately depends on the quality of the manufacturing process and the environmental conditions of transport and storage. These are all inherited attributes.

Organizing the large number of informational links for drugs in the supply chain requires adherence to two concepts:

Data Aggregation is the logical equivalent of item aggregation or assembly. By viewing data within a supply chain as a series of parent – child relationships, track and trace becomes possible.

Data Inheritance is the history of the parent data. To reconstruct the history of an item, each change in form must transfer from parent to child.

Data aggregation reduces the number readings at critical points within the supply chain, making feasible the capture of informational links needed for large-scale drug verification, and track and trace. This becomes evident when dealing with pallet level shipments. Adopting the concept of data aggregation and inheritance allows the opportunity to read a single RFID tag, fixed to a pallet, for specific details about each product on the pallet. If data aggregation were not possible, the RFID tag for each product on the pallet would need to be read, resulting in a great number of reads.

Figure 7-2 shows a visualization of data aggregation for the flow of information between a chemical plant and the manufacturer represented in Figure 7-1. In this case, information flow is in parallel to physical product flow. The information infrastructure built by the RFID center and implemented by EPCGlobal takes advantage of data aggregation and inheritance.

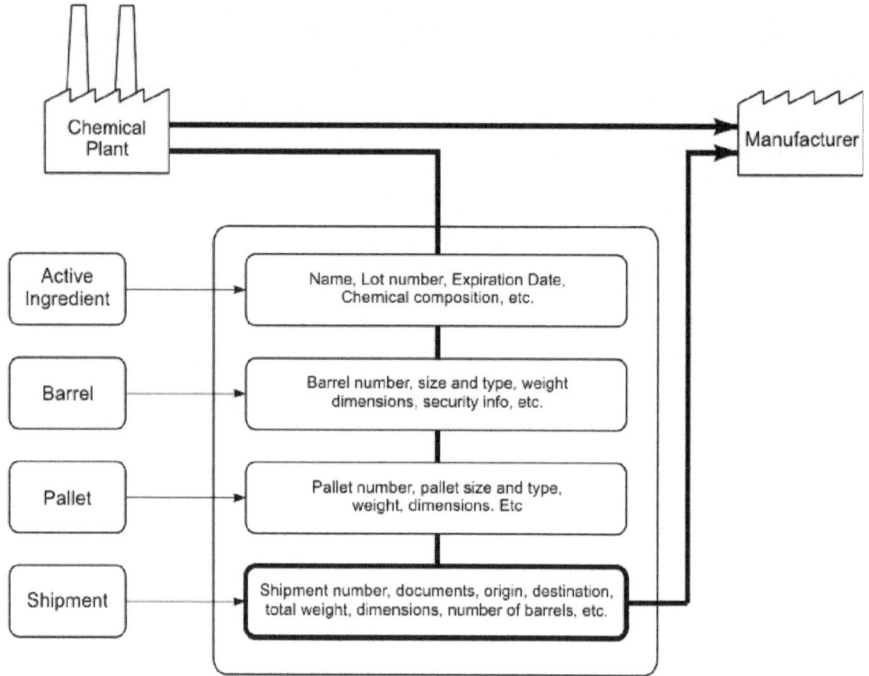

Figure 7-2 – Data Aggregation in the Pharmaceutical Supply Chain

The Flow of Information Within a Pharmaceutical Supply Chain

The underlying success of logistics depends on the flow of information for effective management. In particular, the information flow between two physical locations must synchronize with the parallel flow of goods. *Data pre-positioning* or the use of a *central repository* within the supply chain are important concepts that simplify the synchronization process.[28, 29]

As an example to explain information flows, consider the data needed for shipments between two firms. Finished products in the consumer goods and pharmaceutical industries typically are packed into cases that in turn are loaded onto pallets. A complete shipment consists of a specific number of pallets. Each shipment has an identification number, a bill of lading number, a quantity of pallets, an invoice number, an origin and destination, a driver, and a truck ID number.

Assume that each individual unit, carton, or pallet contains an EPC embedded in an RFID chip. When scanned, the EPC number is linked by the ONS to specific information about the item accessible through the Internet or other type of computer network.[30] To synchronize the flow of both goods and information the shipper must either 1) send in advance (pre-position) complete PML files containing all the information regarding the objects to be received (the *thick file* approach), 2) preposition only EPCs (the *thin file* approach) or 3) write select information to a third party for use by all supply chain partners.

These alternatives represent different ways to share important information within the supply chain. For drug verification, only EPC validation is necessary. This is binary, yes/no information. However, for track and trace purposes transfer of additional information that is dynamic must occur. This information might include:

– Origin
– Destination
– Time stamps
– Company names
– Telemetry information – temperature and humidity

For track and trace, location information is extremely important because it provides 1) the past position of the goods, 2) present position of the goods and 3) the anticipated future position of the goods (assuming a scheduled shipment exists). Time stamps at each location allow the calculation of residence time.

An RFID Based Solution

This section deals with the specific information infrastructure needed for the two alternatives to combat counterfeit; track and trace, and drug authentication. Building on the principles of data aggregation and inheritance, the goal is to gain complete supply chain visibility for detection and control of counterfeit. The outcome of such an approach is an information infrastructure that provides anti-counterfeit measures to deal with several major categories:

- completely fake product
- tampered product
- adulteration
- substitution
- unacceptable status of the product including, expired, discarded, returned, recalled, and samples given by Doctors

Two components of RFID technology form the backbone of supply chain wide visibility. As one group of researchers state: "The key to the RFID architecture is the Electronic Product Code, which extends the granularity of identity data far beyond that which is currently achieved by most barcode systems in use today."[31] In addition, the PML servers, located at each node of the supply chain, and secure Internet based communication combine to provide the primary information handling structure.

Both hardware components, the EPC tags and PML servers, are technologically feasible with significant development having taken place since 2000. In the case of EPC tags, encryption technology, the ability to sense and record temperature and humidity, and the ability to sense certain types of tampering are recent advances that fit nicely with the anti-counterfeit mission. A number of vendors are capable of producing these two infrastructure components in sufficient supplies to meet demands of the pharmaceutical industry. As commercial volumes of production occur, costs will decrease because of learning curve effects. In summary, RFID technology is ready for the challenges of large-scale commercial application.

In preparation for a detailed analysis of drug verification and track and trace, the next section discusses three important databases that will handle the storage and transfer of information.

Database Overview

The proposed RFID technology solution to counterfeit is comprised of three different types of databases with access levels restricted based on business rules defined by supply chain partners. The *Manufacturer Database* contains information about a particular drug and a potential link to third party product registries.[32] The *Central Repository Database* contains the trail of all exchanges for a drug in the supply chain. Administered through a 3rd party, this database contains only select information from manufacturers and distributors needed to secure the supply chain. Finally, the *Local Database* contains information important to the manufacturer or distributor handling the drug. This might include such information as the warehouse location for an individual drug, or management signatures indicating completion of critical procedures. This information is important only to the organization in possession of the drug. Figure 7-3 provides a schematic of all three databases within the pharmaceutical supply chain.

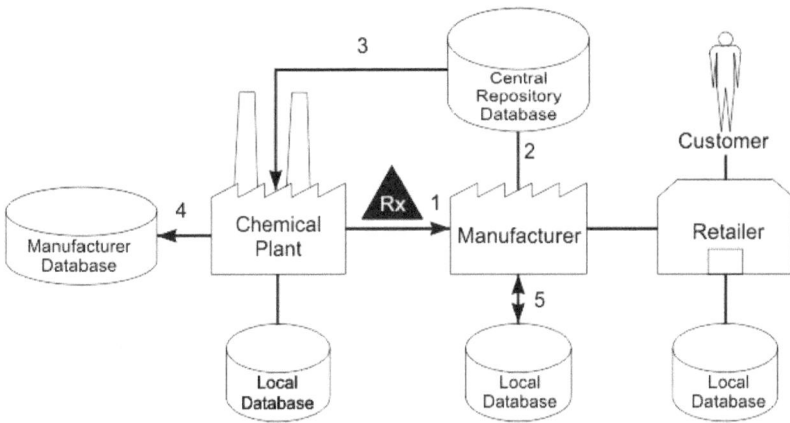

Figure 7-3 – Databases Within the Pharmaceutical Supply Chain

With this supply chain wide view, a broad categorization of information exists:

Static – information that does not change through time (ex. product size and weights)

Semi Dynamic – information that changes with long time intervals (lot numbers based on production run)

Dynamic – information that changes with short time intervals (location, temperature, pressure and humidity readings)

The manufacturer database contains static and semi-dynamic information. In the RFID view, a subset of this database is made available for supply chain wide access using several means, including secure Internet access. In turn, the repository database contains predominantly EPCs and related location and timestamp information obtained from nodes within the supply chain. This information is available to all who have security clearance. Finally, the local database contains dynamic information; some of which might also be written to the repository database.

Counterfeit Detection Within an RFID Enabled Supply Chain

With the RFID infrastructure in place track and trace, and drug verification, becomes possible. The final section of this chapter examines how RFID enables these important pharmaceutical supply chain security measures.

Track and Trace

In 2003, a group of researchers made significant progress in providing the following design for a supply chain wide central repository database (Figure 7-4).[33]

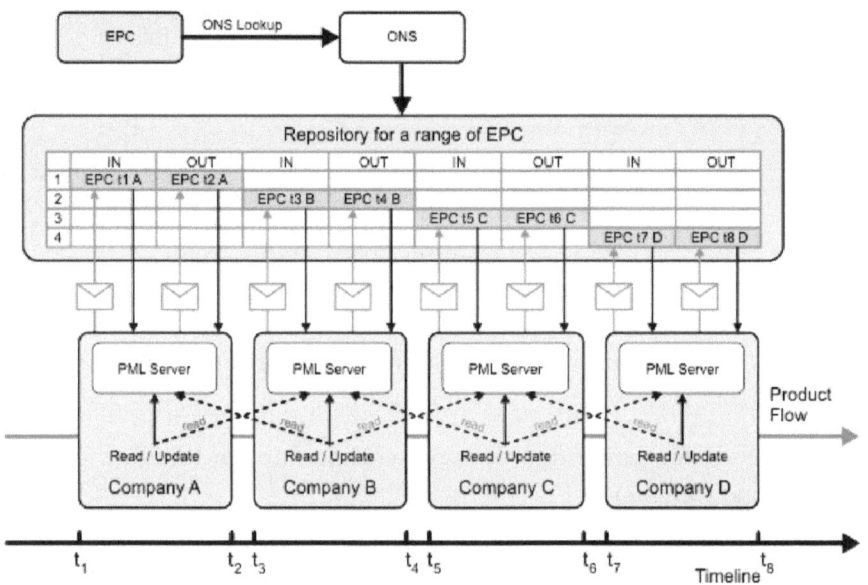

Figure 7-4 – Central Information Repository for the Pharmaceutical Supply Chain

Figure 7-4 shows a supply chain representation with PML Servers as the backbone for routing information. The four step supply chain example in the diagram is analogous to the pharmaceutical supply chain representation given in Figure 7-1, with Company A, B, C, & D corresponding to

1) The Chemical Plant, 2) The Manufacturer, 3) The Wholesaler and 4) The Retail Pharmacy.

In this scenario, information needed for a drug pedigree is written into the central repository from each node in the supply chain. The concepts of aggregation and inheritance reduce the amount of data reported to the repository by associating groups of EPCs with activities that occur at each supply chain node such as storage and shipping. Writing the pedigree information to a single database also increases accessibility to all parties within the supply chain. Real time query is possible.

In contrast to the central repository approach, using a second method, pre-positioning, where information moves sequentially through PML servers as drugs are shipped means that current pedigrees are only available at the farthest point of progress through the supply chain. Information pre-positioning does provide effective tracking capabilities for shipments between two nodes, however, trace information requires following pointers backward through PML servers located in the separate business organizations that make up the pharmaceutical supply chain. This would appear to limit real time access to pedigree information and make supply chain wide telemetric information hard to consolidate.

In summary, the central repository approach proposed provides a more robust solution in creating supply chain wide pedigrees for drugs. With RFID based pedigrees, there will be systematic, timely, and targeted ways to build information for every drug.

Drug Verification

In contrast to building an information infrastructure for pedigrees that depends on a 3rd party repository, drug verification is a simpler case. The informational focus of drug verification centers on the PML files located at the manufacturer. The process of verifying a drug is binary. Either the drug is authentic or it is counterfeit. A thin subset of PML files that contain only valid EPCs, and current status, could be extracted from PML servers and posted for secure internet access. No other information except valid EPCs would be listed. Other supply chain organizations, such as wholesalers and pharmacies could scan the EPCs contained on drugs and compare to valid EPCs posted by the manufacturer. Any discrepancies would be a strong indication of counterfeit.

In this situation, no pedigree information is available, so it is possible that a valid EPC could be mishandled or adulterated as it passes out of the control of the manufacturer. To deal with these issues, RFID could utilize sophisticated anti-tamper tags and integrate with current anti-tampering packaging methods to ensure the physical product is secure.

In summary, drug verification requires a much simpler information infrastructure; however, the approach does not provide as much security. It is possible that implementation of drug verification might be a first step toward obtaining full track and trace capability needed for complete drug pedigrees.

Steps Toward Practical Implementation

In February 2004, the FDA released a report on the future role of RFID technology in the prevention of counterfeit medicines.[34] This was the culmination of several years of public and private research involving the application of technology to improve patient safety. There was also a realization that the United States pharmaceutical supply chain was increasingly vulnerable to counterfeit. Prior to the 2004 report, Mark B. McClellan, former FDA Commissioner had stated the goal of combating counterfeit through "...build[ing] a 21st century system that can better protect consumers against this emerging public health threat."[35]

In the future, the FDA will focus on technologies that allow track and trace of all medicines sold in the United States at the pallet, case, and unit of dosage levels. Though no immediate solution for track and trace has been proposed, the FDA has established a goal of implementing RFID Technology by 2007.

In accomplishing this goal, the FDA also has put forth broad guidelines for implementation after hearing testimony about RFID technology. These guidelines appear in the APPENDIX. This provides an early attempt at establishing the critical factors and activities needed to construct a practical system based on RFID technology for pharmaceutical supply chain security.

It is important to note that the FDA report also mentions other initiatives designed to decrease the chances of counterfeit drugs entering the U.S. pharmaceutical supply chain. The FDA makes clear that in their view RFID technology alone is not the total solution to combating counterfeit drugs. However, the FDA does make a strong declaration in favor of the capabilities of RFID technology by stating that the "use of mass serialization to uniquely identify all drug products intended for use in the United States is the single most powerful tool available to secure the U. S. drug supply."[36]

Both track and trace, and drug verification are feasible through the implementation of RFID technology. While application of RFID technology to combat counterfeit is compelling, several factors offer complexities that must be overcome.

The RFID approach assumes that all drug manufacturers, carriers, wholesalers and pharmacies have the necessary hardware and computing ability to read and process EPC information. It is unrealistic to believe that this capability will occur immediately. However, through merger activity, the number of players in the pharmaceutical industry has decreased. This situation could make the job of implementing an industry wide RFID solution to detect and control counterfeit easier because there are fewer major players.

The RFID approach will have to be fine-tuned in terms of information synchronization among many different supply chain partners to ensure a high level of reliability for pedigree and drug verification information. If a single supply chain partner does not properly handle information, pedigrees might show gaps that would raise counterfeit questions. The RFID approach assumes different entities within the pharmaceutical supply chain can achieve a common level of cooperation in supporting the information infrastructure.

Besides proposed applications in improving track and trace, and drug verification, RFID infrastructure also serves as the foundation for future applications of importance to the health care industry. For example, the Human Genome Project will create greater opportunities for engineering drugs to treat small groups of individuals that suffer from specific illnesses.[37] This marks a shift away from the development and marketing of blockbuster drugs targeted at large population segments.[38]

Traditionally, big pharmaceutical companies have taken a broad approach in drug development "aiming their blunderbusses at big categories in primary care, such as cholesterol reduction, blood pressure control, and depression."[39] By focusing on a handful of drugs targeted at wide segments of the population, the pharmaceutical industry adopted a strategy of mass marketing that significantly influenced revenue growth. A study conducted by a leading consulting firm found that "80% of growth for the 10 biggest drug-makers during the last decade came from the eight or so blockbusters a year launched during the 1990s."[40]

As the pharmaceutical industry enters the 21st century, there are signs of movement away from the blockbuster strategy in drug development and marketing. Currently at least one big pharmaceutical company has taken a lead in creating "personalized medicine" where drug therapies for specific diseases are customized to individual patients based on genetic diagnostic tests that can be performed in the Doctor's office.[41] In addition, other tests can determine an individual patient's ability to metabolize commonly prescribed drugs.[42] It is thought that individual genetic make-up has a great influence on the rate that drugs metabolize and thus the effectiveness of currently prescribed drugs in treating specific diseases such as hypertension.

Although the pharmaceutical industry is in the early stages of personalized medicine, the gradual move toward "designer drugs" and treatment therapies with current prescriptions that are customized to a specific individual's metabolic characteristics will mean big changes to the pharmaceutical supply chain. Designer drugs will be manufactured in small lot sizes on a make-to-order basis. In this environment, logistics and coordination takes a new form as thousands of biotechnology drugs flood the pharmaceutical supply chain. Delivery of these new drugs to the right group of people, at the right time and location, and in the correct quantity presents a challenge that the current logistical system cannot handle effectively. RFID lays the foundation for the management of this not-too-distant complexity through improved supply chain information and visibility, and provides the framework for a safer and securer supply chain.

The Realities of Implementation

Few would argue against creating a safe and secure supply chain for pharmaceuticals distributed within the United States and other countries. Both government and business interests agree that un-adulterated prescription and over-the- counter drugs are fundamental for a healthy pharmaceutical market and the basis for future growth. If counterfeit incidents injure people and undermine the public confidence in the medical system, it will become difficult to introduce new pharmaceutical therapies such as genetic-based drugs that could help many people.

In particular, one instance involving a Long Island, NY teenager who "unknowingly took counterfeit drugs" after a liver transplant, has brought a great deal of attention to the problem.[43, 44] In this case, the counterfeit drugs passed through established channels of distribution, raising further concern about the safety of the United States supply chain. Luckily, the teenager survived, however, there are other accounts of similar circumstances where critically ill people took counterfeit drugs as part of their treatment and eventually died as a result.

While counterfeit drugs have yet to become an acute problem within the United States, isolated accounts of practices employed by counterfeiters have raised universal concern among government agencies and businesses. For example, one pharmaceutical company has reported that counterfeiters "used yellow highway paint to get the right color match for fake painkillers."[45] The yellow paint contained lead and the base for the counterfeit drug contained a combination of boric acid and floor wax.[46]

In another case, a lifestyle drug has been extensively counterfeited through the Internet sales channel.[47] By utilizing the Internet to target customers through spam e-mails and as a means of taking customer orders, production of the counterfeit drugs took place outside of the United States. Though the active ingredient of the drug was present in varying quantities, it is impossible to know the manufacturing conditions under which production occurred or the integrity of the other ingredients contained in the drug. Besides production in un-monitored conditions, this type of counterfeit also violates intellectual property laws, costing billions in lost sales.[48]

As the pharmaceutical industry faces a future that will include more worldwide outsourcing of manufacturing, there will be an even greater need to secure the supply chain to ensure that no counterfeit drugs enter the United States.[49] In achieving this goal, it is important to understand the strengths and drawbacks of various anti-counterfeit approaches involving RFID technology. Though a particular technology might work well in test applications, it is not certain that wide-scale implementation will yield similar results.

Besides technological concerns, the role of industry and government in the practical implementation of RFID is an additional issue of great importance. To date, the FDA has relied upon reports, recommendations, and subtle pressure to encourage implementation within the pharmaceutical industry rather than administrative orders designed to make failure to comply a violation of the law. However, if there is not significant progress in preventing counterfeit, the possibility exists that the FDA will enforce mandatory RFID tagging of drugs, probably sometime after January 2007.

This policy presumes that the FDA has a clear idea of the optimal solution for controlling counterfeit along with the technology needed to achieve practical implementation within the pharmaceutical supply chain. Controlling counterfeit is both a business issue and an issue involving the common good. In achieving the common good, laws and administrative orders drafted with honest intentions sometimes do not achieve planned goals because it is difficult for governments to anticipate the future direction of business. Perhaps the first to make this point, David Hume, a historian and philosopher who lived during the early stages of the industrial revolution in England once noted "...the principles of commerce are much more complicated, and require long experience and deep reflection to be well understood in any state. The real consequence of a law or practice is ... often contrary to first appearances." [50] In this view, the free market is the best determinant of how to control counterfeit in the most cost effective way.

The pharmaceutical companies and wholesalers have significant economic interest in protecting the United States pharmaceutical supply to minimize business risk and enhance the value of their brands. Failure to do so might jeopardize future profits, elevating the financial risk of continuing operations. This raises the prospect of investors discounting pharmaceutical equities for the uncertainty associated with potential episodes of counterfeit drugs. Though no evidence exists to date that Wall Street has discounted pharmaceutical stocks for this reason, the credible threat of this type of disruption exists along with other business risks such as new product development failure, pharmaceutical price controls, and product recalls resulting from unanticipated side effects.

Before analyzing business and governmental issues in the context of implementing RFID technology, the next section provides an overview of the pharmaceutical industries as related to security and the adoption of RFID.

In order to maximize effectiveness, comprehensive anti-counterfeit measures must account for the complete supply chain, from raw materials to final dispensing by pharmacies. Understanding the unique elements of the pharmaceutical supply chain is fundamentally important in providing a basis to evaluate the applicability of RFID.

The Pharmaceutical Supply Chain

The pharmaceutical industry has complex characteristics beyond the need to ensure that counterfeit does not enter the supply chain. Being a regulated industry, a great deal of management attention and capital investment is focused on meeting the Good Manufacturing Practices (GMP) and other regulations as outlined by the Food and Drug Administration (FDA). Customers, including wholesalers, hospitals, and pharmacies, all carry an extensive product line of drugs that require high levels of in-stock availability along with short transit times. An out-of-stock on a critical drug might create a life or death situation.

In many cases, the value of these drugs is very high and special handling conditions such as continuous refrigeration are necessary to guarantee potency. All drugs have expiration dates causing situations where obsolescence becomes an important criterion in manufacturing and logistical decision-making.

After leaving manufacturing plants, pallets of drugs are broken down into individual cases at the first step in the supply chain by wholesalers who eventually distribute small quantities (sometimes at the individual container level) to thousands of pharmacies. This is in contrast to the food and consumer goods industries where full pallets of fast moving goods are often shipped directly to retail outlets. Given the special handling demands of the pharmaceutical industry, it is not surprising that logistical costs are high relative to other manufacturing industries.[51]

The Pharmaceutical Manufacturing Process

Taking a broader view, the pharmaceutical industry is part of a category called process-oriented manufacturing. This form of manufacturing is characterized by "mixing, separating, forming, and chemical reactions."[52] The bills-of-material used for production are flat, consisting of only several lev-

els. This is in contrast to heavy manufacturing, such as production of loco-
motives or airplanes, where the BOM is complex and deep containing a
number of different levels.[53]

In conjunction with flat bills-of-materials, product families in pharmaceutical manufacturing take on the additional trait of a "V" structure. Generally,
V-shaped product families have the following common characteristics: [54]

1. The number of end items is large compared to the number of raw materials.
2. All end items sold by the plant are processed in essentially the same
   way.
3. The equipment is generally capital intensive and highly specialized.

With this manufacturing structure, inbound raw materials required for
manufacturing are subject to lot control and quality assurance checks.
Because of the relatively simple BOM structure and manufacturing process
within the pharmaceutical industry, raw material lots can be traced to specific production runs of end items. This is useful information in situations
where a manufacturing process failure or a contaminated raw material
leads to a recall of finished goods. Most Enterprise Resource Planning
(ERP) systems used in the process industries provide some means of keeping a history for each lot of inbound raw material.

Given the established practice of inbound raw material lot control within
the pharmaceutical industry, there is little immediate need for using RFID
tags as a means of tracking and tracing. Though tags will ultimately appear
on raw material packaging, the current practice of lot control is sufficient to
ensure security.

The FDA is concentrating its efforts on track and trace capabilities for finished goods traveling through the supply chain. In the context of these
efforts, the main supply chain players involved in establishing security are
the manufacturers and wholesalers. Pharmacies, hospital or retail, play a
role only in being the end-point for receiving and dispensing drugs.
Though there are some cases where pharmacists have diluted drugs for
personal gain,[55] the main source of concern for counterfeit entry into the
supply chain involves the links between manufacturers and wholesalers.

Since the problem of counterfeit drugs is interwoven within the industrial organization of the pharmaceutical industry, it becomes important to understand the practices that govern the industry.

Industrial Structure of Wholesalers and Manufacturers

The dominant feature of the pharmaceutical supply chain is a two-tier structure of wholesalers. While three major wholesalers account for more than 90 percent of the drugs distributed in the United States, there are thousands of smaller, second tier wholesalers located thoughout the country.[56] In Florida alone, there are 1,399 registered second tier wholesalers, one for every three pharmacies located within the state.[57]

The three large wholesalers are huge companies. All are included in the Fortune 50.[58] Even though sales for these companies are in the billions of dollars, profit margins are slim, often less than 2 percent of revenue.[59]

With thin profit margins, wholesalers have engaged in a number of activities to boost marginal profits. These include arbitrage operations intended to take advantage of price discrepancies within the market, particularly between primary and secondary wholesalers.

These price discrepancies develop because of an imbalance in supply and demand. One reason that imbalances occur is the historical practice of "channel stuffing" by pharmaceutical manufacturers as a means of increasing sales at the end of a quarter or a fiscal year.[60] By announcing price increases in advance to wholesalers, pharmaceutical manufacturers encourage forward buying before the price increase takes effect.

The result of this practice is an accumulation of surplus inventory at the wholesalers. Aware of the approximate timing of manufacturer price increases, wholesalers go one-step further by having sophisticated logistical models that calculate the exact amount of a forward purchase given inventory carrying and storage costs. By one estimate, approximately 40 percent of wholesaler profits come from speculative buying.[61]

The disadvantage of speculative buying is that when large inventories accumulate at the wholesalers there is an increased risk that the drugs will pass expiration dates before final sale to pharmacies. To reduce this risk, primary wholesalers unload inventory that is near expiration to secondary wholesalers at a deep discount. In turn, the secondary wholesalers contact retail pharmacies, hospitals, and nursing homes offering drugs at reduced prices, usually on a one-time-only purchase basis. Likewise, retailers sometimes sell excess inventory to secondary wholesalers to avoid problems with expired product.

Since there are such a large number of secondary wholesalers that engage in re-selling practices, it is extremely hard to trace the path of drugs to pharmacies or to monitor storage conditions. Often secondary wholesalers are small operations consisting of little more than a refrigerator for storage.[62]

Given an active secondary market and complex product flows, there are many opportunities to introduce counterfeit drugs into the United States supply chain through secondary wholesalers. Making matters even more complex, there have been unproven accusations that primary wholesalers have illegally diverted drugs as a means of receiving double rebates from manufacturers for the same purchase.

An allegation put forth in the national news media claimed that a particular primary wholesaler sold drugs to an institutional pharmacy and then collected a rebate from the manufacturer according to standard terms. Based on prior agreement with the primary wholesaler, the institutional pharmacy then sold the drugs to a designated secondary wholesaler. The primary wholesaler in turn repurchased the same drugs from the secondary wholesaler and "cycled the drugs back through the system, each time selling them to an institutional pharmacy to collect an additional rebate from the drug company."[63]

This type of diversion, while offering even more opportunities for counterfeit to enter the supply chain, also takes advantage of the manufacturer's inability to identify drugs uniquely once shipments have left manufacturing plants. Unique identification is a fundamental capability needed to accurately match rebate claims by wholesalers to specific shipments made by manufacturers. With current bar code identification technology, there are few ways that manufacturers can detect double rebates claimed by the primary wholesaler for the same drug.

In summary, the concentration of wholesalers in the middle of the supply chain creates a situation where primary wholesalers command a great deal of market power over manufacturers and pharmacies alike. Secondary wholesalers play a lesser role in making a market for various drugs, usually when some urgency is involved in moving drugs quickly though the supply chain at reduced prices.

The perspective of the pharmaceutical supply chain is an important fundamental in understanding the issues of implementing RFID. The next section provides additional details concerning the use of RFID to control counterfeit.

Basic Implementation Issues

The initial research and development of RFID technology conducted at the MIT Auto-ID Center concentrated on improvements for the consumer goods industry, in terms of decreased out-of-stocks at retailers and general inventory reduction through greater supply chain visibility. Counterfeit control was not an initial research consideration. Early on, an influential group of consumer goods companies, including Proctor & Gamble and Unilever, and retailers such as Wal-Mart and Target, recognized the potential of RFID for operational improvements. In many ways, the technology represented the next generation of bar code. Consensus agreement existed that bar codes were hugely successful since implementation during the 1970s. Much enthusiasm existed for full development of RFID within the consumer goods industry.

It was later that research uncovered the alternative of using RFID technology for counterfeit control. The first publication by the MIT Auto-ID Center concentrating on RFID as an anti-counterfeit measure for the pharmaceutical industry was in 2003,[64] though other authors from industry have claimed they proposed similar ideas as early as 2000.[65] While a number of pilot tests proved commercial viability of RFID within the consumer goods industry,[66, 67] the process of full-scale deployment is still in the early stages. With no proven benchmarks, it is not clear whether the pharmaceutical industry will take a similar implementation path given supply chain characteristics that differ from those found in the consumer goods industry.

Because of the urgency that the counterfeit problem presents, some experts predict, "the pharmaceutical industry is going to move faster than other industries ..." in adopting RFID technology.[68] Others maintain that by 2007 the use of RFID by drug makers will surpass applications by consumer packaged goods companies.[69]

However, this estimate is based on benefits derived from improved inventory management, better product recall capabilities, and enhanced patient safety in addition to counterfeit control. Since RFID is a flexible system that can address many different concerns, it may very well be the case that economic justification must take place on a number of different levels within the pharmaceutical supply chain.

Efforts to Organize Implementation

To coordinate RFID implementation and to address the special needs of the pharmaceutical industry, EPCGlobal has established a Healthcare Life Sciences Business Action Group (HLS BAG).[70] The group consists of a number of working committees, including: information, policy, process, research and development, strategic planning, technology, applied tag performance. The number and scope of the working committees underscores the practical complexity in applying RFID technology to control counterfeit in the pharmaceutical industry. Supporting these efforts, several consulting companies have also conducted focused pilot projects using a consortium approach.[71]

The overall goal of all this work reduces to simple implementation objectives in the following general areas: education, performance, standards, cost and safety.[72] It is only through a comprehensive understanding of the theory of RFID, including tags and readers, combined with experimentation and development that practitioners will realize the full potential of RFID.[73]

As a starting point to establish a foundation for implementation analysis and practice, several factors will play an important role regarding the application of RFID in the pharmaceutical supply chain.[74] The balance of this section details these factors.

Reading RFID Tags on a Large Scale

Overall, the tag read rate is perhaps one of the most significant issues facing the pharmaceutical industry as the FDA deadline approaches for implementing an electronic based system to control counterfeit. The approach in dealing with tag readability issues is a source of significant divergence between the consumer goods and pharmaceutical industries.

In the case of the consumer good industry, system based methods exist to compensate for the current lack of 100% read rates. Some companies are considering adopting an "inferred read" approach to deal with this fundamental problem.

However, this approach assumes that the aggregation is always intact i.e. all items are in a case or all cases are on a pallet. This is a disadvantage when data is needed for such things as a drug tracing through the supply chain where the EPC code for each package must be linked to previous shipments.

If a high read rate for tags placed on each package of drugs is not possible, the alternative is to enter tracking and tracing information manually by EPC code. With the large volume of drugs moving through the supply chain, even partial manual entry of information needed for tracking and tracing is overwhelming. An EPC used in the pharmaceutical industry could contain 31 or more digits.[75]

For these reasons, inferred reads are probably not effective in the pharmaceutical industry where 100% read rates are an important element of automatically generating tracking and tracing information for every step of the supply chain.

Choice of Radio Frequency Standard

An important variable in achieving successful reads for tags is the frequency employed. Specific frequencies work better in certain situations such as scanning tags placed on objects that are metal or contain liquids. It is possible that the final implementation of RFID in the pharmaceutical industry will mean different tags, with different frequencies, working together.

For example, one frequency could be used for tags placed on individual packages and a different frequency for tags places on cases and pallets. To understand what works best, the pharmaceutical industry will need to go through a process of trial and error in testing tags of different frequencies before deciding the best mix of frequencies to use. An important part of this process is full scientific and engineering knowledge of the properties associated with electromagnetic fields and RFID.

Product Stability

Before large-scale application of RFID can take place in the pharmaceutical supply chain, testing must occur concerning the impact of electromagnetic fields on the stability of drugs. Increasingly, pharmaceutical products are becoming highly bioengineered molecules that are inherently unstable in nature. For many of these medicines, overexposure to heat or light can degrade chemical bonds and render the drug ineffective. Though the chances are low, the additional energy transmitted from readers in the form of magnetic fields might be enough to break critical chemical bonds within medicines.

To date, there is no direct evidence that the levels of energy associated with prolonged exposure in an RFID system, such as in a warehouse, are high enough to cause product quality problems. However, testing must take place to provide empirical evidence for conclusive proof before RFID technology can move forward as an aid to supply chain security and other functions such as improved inventory control. Though the amount of heating that takes place when a drug is in the presence of the reader field is minimal, other non-thermal effects could break chemical bonds within molecules. Biologics, such as vaccines, are at the greatest risk for breakdown. However, the entire question of product degradation still requires more study.[76]

Electromagnetic Compatibility

Hospitals and medical clinics are filled with complex electronic equipment used to maintain and save lives. Specific frequencies of radio waves that are associated with RFID tags might negatively affect the operation of some types of medical equipment. For example, the use of cell phones within hospitals has been banned because of fears that the signal might interfere with medical equipment.[77] Potential interference must be completely understood before full-scale implementation of RFID can take place.

The Code Structure of the EPC System

Identifying an individual item within the supply chain by using a mass serialized identifier is the corner stone of the EPC system. However, less than full agreement exists within industry concerning the essential elements of the code structure

There are two approaches to code structure: 1) use the EPC as a substitute for existing codes, such as the National Drug Code (NDC), or 2) create a new structure by nesting previous codes, such as the NDC, within the EPC. The decision on which approach to adopt has worldwide implications because each country has a different code structure for the current identification of drugs. With outsourcing of manufacturing occurring on a worldwide basis, this will become an important issue for consensus.

ONS Constraint

One of the important design characteristics of the EPC system is the storage of a small amount of data on the tag, (the EPC, a 96 bit serial number). Storage of relevant information about the object (drug) resides on servers that can be accessed through the Internet. The mechanism to link the EPC to a server location is the Object Naming Service (ONS). One fundamental problem of the current release of ONS is that it does not specify the linking mechanism at an individual item level. This means that a drug can be traced to an individual manufacturer, however, no link through the EPC exists to specific information about the drug located on the manufacturers' servers. This offers significant drawbacks in authentication of specific drugs and the generation of tracing information that accounts for all drugs within a lot.

Privacy Issues

With the potential of linking specific drugs to an individual through a large-scale system, maintaining privacy is a significant issue for further research. The direction of this research will probably move toward various encryption technologies to cipher the link between a specific drug and an individual.

Cost of Tags

Under ideal situations, end-item tagging would occur for all pharmaceutical products at the package level. With this scenario, all packages contained within a case, along with all cases and pallets would contain tags. This would create demand for billions of tags.

At these volume levels, early research projected that cost would approach 5 cents per tag after industry wide adoption of RFID.[78] However, recent projections show the 5 cent tag to be a distant goal beyond 2008. In the near term, tag costs should stabilize at 16 cents per unit.[79]

At this cost, it is unlikely that end item tagging will take place for a wide range of pharmaceuticals. In one early case, limited testing involving a single drug resulted in an initial technological infrastructure cost of $2 million along with 50 cents for each tag.[80] Although prices for RFID technology should decrease with time, early adopters face significant costs.

Redundancy

A final issue worth considering involves the need to build a redundant system for securing the pharmaceutical supply chain. The dispensing of drugs is sometimes a life or death situation. If verification of the authenticity of a drug is not possible because of system failure, then lives could be lost. Before full implementation of RFID technology can take place in the pharmaceutical industry, significant research must take place to ensure system reliability never becomes a factor in slowing the administration of drugs.

In some cases, suppliers are adding bar codes to RFID tags as a measure of redundancy in the case of tag failure. Along with this approach, suppliers are also offering hand-held readers that can read both bar codes and tags.[81]

Choosing an Industry-Wide Architecture

Beyond the basic issues of implementing RFID are greater questions concerning the proper industry-wide information technology infrastructure to use for controlling counterfeit. Choosing an architecture will have significant impact of the overall success of RFID in practice.

As previously mentioned, there are two basic approaches: 1) track and trace to create a pedigree that will accompany each drug throughout the supply chain, and 2) drug verification where a manufacturer database can be queried remotely to check the authenticity of a drug.

In the case of track and trace, data on individual drugs could be handed-off through the supply chain, creating a pedigree. This is sometimes called the "daisy chain" approach because of the sequential nature of information flows between entities within supply chain.[82] While these sequential information flows would be handled electronically, passing from wholesaler to wholesaler, it is also possible to write data to a central repository, assessable from anywhere within the supply chain. With the central repository approach, pedigrees would be updated for each move through the supply chain.

In contrast, the verification approach is independent of a drug's location within the supply chain and is similar in architecture to commercial credit card processing.[83] With this approach, a pedigree does not trace with the product movement. Rather, at the end of the supply chain just before dispensing, the EPC is verified against a database of valid EPCs held either at the manufacturer or centrally.

While there has been no clear trend on which supply chain architecture the pharmaceutical industry will adopt, there have been some interesting developments that will point the way for future anti-counterfeit systems. In one case, several technology firms have collaborated to develop an end-item verification system using 13.56 MHz RFID tags. The system uses public key infrastructure (PKI), and provides both track and trace, and verification in two steps.[84]

In the first phase, tags and the PKI would be used to provide an off network method of verifying the authenticity of drugs at dispensing (the pharmacy). This approach does not require manufacturers and wholesalers to do a full implementation of RFID technology within the pharmaceutical supply chain. The benefit of this approach is a minimum investment in infrastructure.

The second phase combines the PKI technology with e-pedigree software. By using specially developed readers at shipping and receiving points, the verification process could be duplicated many times as a product moves through the supply chain. A single time stamp would accompany each verification. This would provide documentation of all of the steps from manufacturers to pharmacies, fulfilling the pedigree requirements established by select states and the federal government.

Regardless of the architecture chosen by industry, the critical issue remains concerning the reliability of tag reads. If a tag fails to read properly, there is no data input and no way of establishing a pedigree or verifying a drug is authentic. Because of the uncertainty of knowing when in the future tag read rates will reach the standard establishing by bar code technology, many industrial and software/hardware firms are looking to other methods to control counterfeit.

Alternatives to Using RFID Tags

Using RFID tags as a means of counterfeit control takes advantage of the EPC code to identify an object uniquely. While RFID tags are one means of unique identification, other alternatives exist that involve encoding a serial number into a 2 dimensional (2-D) bar code. The advantage of this approach is that 2-D bar codes are thought to have better read rates as compared to RFID tags. However, it is also true that 2-D bar codes have never achieved the reliability of convential bar codes, so there remains the question of how much is gained through using the 2-D bar code as an alternative to RFID. Another drawback is that the serialization approach taken with 2-D bar codes does not utilize the EPC numbering system, although there are no technical reasons that restrict the EPC to RFID tags.[85]

During the transition period between bar codes and RFID technology, the use of 2-D bar codes is drawing attention as an attractive alternative.[86] To highlight the point of 2-D bar codes being a transitional technology, one vendor representative comments "… we are giving people the ability to do today what RFID will allow them to do 10 years from now."[87]

Besides the use of 2-D bar codes, there are indications that the pharmaceutical industry is responding to the counterfeit problem by using means other than identification technology to control counterfeit.

In at least one case, a primary wholesaler has announced plans to eliminate its drug trading business as a means of reducing the chance of counterfeit drugs entering the supply chain.[88] Following this lead, a major drug retailer with 5,400 stores announced that it will no longer purchase pharmaceuticals from the secondary market.[89]

Manufacturers have also taken steps to reduce the chances of counterfeit entering pharmaceutical supply chain by limiting the amount sold to wholesalers to one month of inventory."[90] In addition, drug manufacturers have developed special software that detects instances of speculative buying by wholesalers.

Other Anti-counterfeit Applications

Pharmaceuticals are not the only products subject to counterfeit. The World Customs Organization has estimated that counterfeit equals 5% – 7% of the global merchandise trade or about $512 billion in lost sales for 2004.[91] Just about any brand name merchandise can be counterfeited. There is even a documented case of a counterfeit elevator system located in a high-rise building that "… stopped between floors."[92]

The abilities of counterfeiters have become sophisticated with expertise in shipping, warehousing, and other logistics activities needed to infiltrate existing supply chains. Since the profits from counterfeit merchandise are lucrative, large criminal organizations are becoming involved in the practice. Some believe that at least one Asian government has a policy of organized counterfeit operations directed at both Western and Asian goods that grosses as much as $500 million per year.[93]

Because of improvements in technology, low value products are also subject to counterfeit. The Subway restaurant chain announced discontinuation of decades-old free sandwich program because counterfeiters were reproducing proof-of-purchase stamps. Thousands of these stamps appeared for sale on online auction sites.[94]

With increasing globalization, and greater opportunities for counterfeit to enter legitimate supply chains, the role of RFID in terms of tracking, tracing, and verification should increase in importance. Testing of anti-counterfeit measures in the pharmaceutical industry, an important aspect of patient safety and homeland security, should pave the way for wider applications in other industries. Though there exists a great deal of work to make RFID an everyday reality, the prospects of reducing counterfeit of pharmaceuticals and general merchandise offers great incentives that are hard for any industrial firm to ignore.

Appendix [95]

1 Business Steps for Industry

Each industry stakeholder interested in implementing RFID would benefit from the following steps:

- Create an internal team focused on the adoption of mass serialization and use of RFID technology;
- Perform internal feasibility studies to gain experience with mass serialization and RFID technology and to identify internal business issues requiring resolution;
- Perform external pilot studies with stakeholders across the supply chain to gain experience using mass serialization and RFID and to identify opportunities, barriers and external business issues associated with them;
- Develop policy and a business case for the use of mass serialization and RFID;
- Cooperate and work with other stakeholders and government agencies to develop infrastructure and information systems to use with mass serialization of pallets, cases, and packages of drugs;
- Participate on standard setting groups developing technical standards and business rules for use of mass serialization and RFID;
- Work with government agencies and other members of the supply chain to identify and address regulatory and economic issues that could delay the adoption of mass serialization and RFID; and
- Educate other members of the supply chain and government agencies about mass serialization and RFID.

To the extent possible, it would be most useful for interested firms to perform these actions concurrently. For example, standards development requires knowledge gained from feasibility studies in order to move forward, and vice versa.

2 Standards Setting Issues

Any effort to develop standards for mass serialization of pallets, cases, and packages would be most effective if it addressed the following issues:

- Minimum Information Requirements for the serial number – in the case of RFID tags this means containing a mass serialization code that uniquely identifies the object to which it is attached (e. g., minimum of 96 bits of information);
- Communication protocol standards – in the case of RFID this means standard protocols for interrogating and reading tags;
- Reader Requirements – Readers of mass serialization codes should be interoperable (e. g., readers must use protocols that allow them to read multiple classes of tags or bar codes, as applicable) and should be able to automatically upgrade software over an information network;
- Pedigree requirements – this means that databases containing transaction information should be compatible (e.g., format, mark-up language);
- Information Network Requirements 1. Database Structure (e .g., centralized vs. distributive) 2. Data ownership 3. Data access (to meet business, track and trace, and recall needs) 4. Data Access controls to assure information security;
- Software Requirements – all applications should be compatible and compliant to assure global interoperability; and
- Best use of Frequencies – (e. g., 13.56 megahertz on packages and 915 megahertz on cases and pallets due to interference and read range issues).

CHAPTER 8

Medical Devices:
Smart Healthcare Infrastructure

The maintenance of health and the treatment of disease are clinical sciences that rely upon skilled observation and testing. This generates a large amount of data in the form of patient records, which are often vital in making life or death decisions. In spite of the importance of data and information, the healthcare industry "lags behind other industries in using technology to store and retrieve data, to the detriment of doctors and patients."[1]

Employing a large amount of expensive assets, the practice of medicine consumes about fifteen percent of the US gross national product each year with costs continually growing faster than the rate of inflation.[2] Given demographic shifts within the US during the next twenty years, costs will increase even more as the number of people reaching old age goes up.[3]

In the context of these developments, the EPCglobal Network and RFID technology are likely to play an important role as the medical system responds to the challenges of reducing costs. One study estimates that the market size for RFID products within the healthcare industry will be $8.8 billion by 2010.[4] Polls show that most Americans support new technologies to improve the quality and reduce the cost of medical care. For example, more than 83 percent expressed a willingness to adopt home monitoring and testing, including basic medical procedures.[5]

Large corporations are watching these trends closely. General Electric has a program to deliver "personalized healthcare" that will tailor treatments to individual patents. With an emphasis on early diagnosis, the company anticipates rapid growth in this area. Already, GE Healthcare is "nearly as big as the GE Transportation unit, which includes aircraft engines."[6]

While the application of RFID in the medical industry is in its early stages, there are several emerging areas worth examination. The ability to provide continuity of care, continuous patient monitoring, shared yet secure medical records, valid and accurate medical dosages, medical equipment tracking, and improved information display and communication, are some of the opportunities provided by the proposed infrastructure put forth by EPCglobal. RFID technology has the potential to increase the effectiveness, reduce costs, and ensure the reliability of healthcare services.

An Intelligent Infrastructure for Healthcare

The vision of the EPCglobal Network and RFID technology is to provide continual access to the identity, location, and the state of physical objects. This has a number of applications for medical practice and healthcare in general.

Item Identification and Tracking

The most apparent application of RFID technology to healthcare is its use in the identification and tracking of medical products and devices. From larger equipment, such as wheel chairs, gurneys, incubators and anesthesia carts; to diagnostic tools, such as portable ultrasound, endoscopes, aspirators, insufflators and defibrillators; and to small products, such as surgical instruments, medications, syringes, clothing, and dressings; real-time, automatic tracking will provide a direct increase in clinical efficiency while also reducing cost. For example, "If a hospital just knew what medical equipment it had, and where it had it, it wouldn't buy multiple pieces of equipment that go unused."[7]

Using RFID as a means of tracking expensive assets is not a new idea. Previous surveys indicate that it is a leading application in industries outside of healthcare.[8] Several companies sell tracking systems that utilize active tags operating at 433 MHz, and in some cases Wi-Fi networks.[9, 10] Based on experience in installing tracking systems, one author projects a 1-3% decrease in operating costs through reduced loss, improved tracking of maintenance, and protection from theft, fraud or damage.[11] Since most of medical assets are of high value, the cost of the tag is a low percentage of the value of the asset. This makes the medical industry one of the best candidates for early RFID adoption.

Even the electronic tagging and tracking of medical records, charts, and films may have near-term benefits pending eventual conversion to electronic formats. All of these applications should improve inventory management, equipment utilization and tracking, and theft prevention.

Medical Instrument Tracking

Besides tracking assets or medical records, RFID has other specific applications in tracking. Surgery involves a number of different instruments, some very small. Various procedures require specific sequences of insertion and removal. Though a checklist for removal of instruments exists, it is vitally important to "ensure they're not left inside a sewn-up body." [12] Placing small, passive tags on instruments would enable a surgeon to perform a scan of the patient after completion of surgery to be positive that no foreign objects remain.

Beyond the function of tracking during surgery, there are other important reasons for tagging medical instruments. After use, surgical instruments must be properly scrubbed, autoclaved, and packaged in preparation for the next medical procedure. Electronic tags on the instrument and tag readers on the instrument trays, in the sterilization chamber and in the storage cabinets can locate instruments and validate proper cleaning. A single example shows why this is important.

In 2004, nearly 4,000 patients were exposed to medical instruments that were not properly cleaned.[13] The specific case involved an unforeseen situation where the medical instruments were mistakenly cleaned in hydraulic fluid instead of soap prior to sterilization.

The mishap occurred because elevator workers "drained hydraulic fluid into empty soap containers without changing the labels." [14] Other workers then used what they thought was soap to clean the instruments.

With RFID enabled systems, medical instruments and cleaning agents can be tracked simultaneously, creating "fail-safe" validation. With sophisticated middleware, the EPC associated with specific medical instruments could be associated with the EPC on the container of cleaning agents used.

This type of linking provides insights into cause and effect that are important in addressing complex problems like the causes of infection and the effectiveness of products and procedures.

Monitoring Radioactive Isotopes

A final area where tracking is extremely important involves potentially hazardous materials. Tracking radioactive isotopes in a hospital is currently a major administrative and security burden for the medical staff. The radiation safety officer must monitor, control, and record the interaction and location of radioactive materials for every step in the medical process – from storage to transport to administration to disposal. Automated monitoring of these controlled materials would greatly relieve bureaucratic complexity, as well as increasing safety and security.

In all medical applications, having a standard for unique identification, which is the goal of EPCglobal, is a fundamental part of creating computer systems to better utilize assets and to increase safety. The general idea of linking physical objects to a network has great potential to change the nature of medicine through new forms of data and interaction previously unavailable.

Continuity of Care

The general approach of the EPCglobal Network in using unique identification to create a base for an integrated, intelligent infrastructure might have the greatest impact on medical practice and healthcare information networks. This is especially true given the traditional organization of the healthcare industry.

Medicine, as it exists today, segments care into specific domains ranging from the intensive care unit in a hospital to the medicine chest in the home. In every case, these domains are separate from one another with few if any means of integration.

For example, in a modern hospital, there are many different departments, each with varying degrees of monitoring and care, across a wide range of specialties. Within each department (dermatology, medicine, neurology, obstetrics, orthopedics, pathology, pediatrics, radiology, surgery, etc.) patient records, registration, monitoring and display information do not exist in a common format.

Beyond physician offices and hospitals, there are many other places where medical care takes place. These include emergency personnel (police, fire, and ambulatory services), their vehicles and facilities; and assisted living and nursing facilities, as well as individual homes, businesses, and public transportation systems.

The disconnection between patient information and the healthcare provider is even more of an issue across international borders. Language barriers, standards of care, recording methods, units of measure, communications infrastructure, and time zones, all contribute to a nearly impenetrable wall between physician and information about a patient. At best, this may simply inhibit or delay treatment. At worst, lack of relevant information might be life threatening.

In cases where a lack of complete information exists, which is true in almost every medical instance, physicians and other medical personnel must rely on physical examination, follow-up observation, judgment, and experience to make decisions concerning the correct course of medical care. An important objective of the EPCglobal Network and RFID technology is to improve this situation through employing a common standard for unique identification of physical objects. With a standardized approach, a base is set for information to move freely from one medical domain to another. This enables such things as remote medicine where physicians can administer treatment at a distance without direct contact between a physician and the patient. In addition, common standards also allow new software development for a host of different applications including user interfaces, data management tools, and archival systems.

Patient Identification and Location

The patient is the most important person in a hospital or clinic. All efforts focus on his or her treatment, recovery and eventual discharge. Just as the medical ID bracelet has for years provided positive identification of a patient, an RFID tag containing the EPC might provide accurate, automatic, and real-time identification of the person under medical care. Through the EPCglobal Network Infrastructure, such an electronic identification system would allow all patient information to "follow" them wherever they may go. This has great potential to integrate data and information from many different disciplines and to help in situations where a patient is incapacitated, or in situations where communication is not possible such as with an infant.

Employing an RFID approach, the physician could know patient location in real time. Although this is not necessary for all circumstances, tracking patient movement throughout a medical facility for the critically ill would be very helpful.

With the standards that EPCglobal puts forth, it becomes possible to develop sophisticated network interface devices such as a personal digital assistant (PDA) or bedside display, providing access to the most up-to-date patient data.

From the point of view of the nursing function, knowing the current location and history of movement for a patient could be extremely helpful in providing efficient care and treatment. Given hospital environments where patients might be mobile, the value of real-time location reduces search time, a significant cost within large medical facilities.

In addition, information about patient location is a fundamental input to advanced scheduling systems capable of maximizing resource utilization. For the hospital administrator, this information is critical in sizing resources to match patient demand for the near and long-term.

Patient identification also has implications concerning medical insurance. With unique identification and RFID enabled medical facilities, it becomes possible to keep track and verify the interaction between physicians and patients. This type of automated system has great potential to eliminate billing errors and fraud.

In the extreme case, it is now possible to implant a tag into a person that contains important medical information in the event that a patient losses consciousness.[15] The chip is the size of a grain of rice and can provide a link using the EPC to important information such as allergic reactions to medicine.

Drug Validation

With the large number of prescription drugs available for treatment, a great deal of potential exists for administering the wrong drug. In the effort to reduce costs, medical personal are under constant time pressure. Similar drug names can often be confused.

In particular, the FDA has issued a warning concerning two drugs with similar names; "Zyprexa, an antipsychotic drug made by Eli Lilly & Co.

and Zyrtec, an antihistamine made by Pfizer, used to treat allergies."[16] Complicating matters, these drugs are often stored near each other on pharmacy shelves, and have the same dosage (once per day).[17] Since 1996, there have been at least 79 reports of prescribing errors for these two drugs.[18] As might be expected, the situation where Zyrtec is prescribed or mistakenly dispensed by a pharmacy instead of Zyprexa can lead to serious consequences.

Assuming the patient wears an RFID tag while in the hospital, a simple drug record and validation process could be performed automatically to alert for possible errors. While current safeguards minimize mistakes, an automatic system (if properly administered) could provide reassurance, and eliminate rare, but costly medical errors. Similar types of verification systems could also be installed in pharmacies.

Compliance

In addition to the proper administration of drugs, there is also the increasingly important issue of compliance. To measure the response of a patient to a particular drug, physicians presume that the drug is taken in the proper dosage and at the correct time as prescribed. While in a hospital or nursing home setting it is more likely that compliance takes place, it is far more likely that doses will be missed if a patient is at home, especially if they are alone.

Though tags placed on pharmaceutical containers, along with readers located in the medicine cabinet, could identify removal of the container there still would be no actual verification that the dose was taken. While other approaches such as placing removable tags on individual pills have been proposed, there remain considerable safety issues to overcome. In the future, it might be possible to print a tag onto an individual tablet that can be safely ingested. This would provide a means for determining accurate compliance.

Sensors and Telemetry

Measuring the vital states of a patient is the first priority for doctors, nurses, and technicians. From the basic "ABC's" (Airway, Breathing and Circulation) to sophisticated blood chemical analysis and three-dimensional medical imaging, measuring, understanding, and recording patient data is an ongoing activity in the healthcare facility.

Some clinical states, such as general anatomy or chronic lesions, change slowly over time. Others change more rapidly. Examples include glucose levels, blood gas, or temperature. Still others change almost instantaneously. These include cardiac electrical state, muscle activation, or nerve conduction. Sensor systems exist for measuring many physiological states such as heart rate, electrocardiogram (EKG), respiration, blood oxygen saturation, temperature, body position, muscle activity and others, as well as analytic systems for measuring states of recovered tissue and fluid samples.

What is common to all these systems in the measurement and recording of data values and associated information, such as the date and time of measurement, characteristics of the measuring device, and the particulars of the patient. All of this data can be linked to the EPC through the underlying infrastructure that comprises the EPCglobal Network. This provides a universal means of referencing data and information associated with a specific patient.

The active tags currently available have the potential to provide telemetry in addition to unique identification. Active tags equipped with sensors could be adhered to the patient, providing continuous measurement of physiological variables. This measurement could be transmitted wirelessly to the hospital reader network, for communication to the appropriate hospital staff. Furthermore, since the tags would comply with the proposed open standards of EPCglobal, their transmission would work equally well in the hospital, clinic, emergency transport or home.

There has also been research conducted into using passive sensors for medical data. In one case, a small passive tag and sensor can monitor blood glucose levels for people with diabetes.[19] By waving an arm or hand containing the implanted sensor close to a reader, the blood sugar levels would automatically be recorded. This eliminates the need to draw blood for testing.

The Future

While much of the focus of RFID technology has concentrated on supply chain management, there remain significant opportunities for implementing the EPCglobal network in healthcare. This chapter details just a few applications that are likely to become commonplace in hospitals, clinics, and even the home. The first step in gaining value from RFID is the refinement of a standard to identify physical objects. With an established means of unique identification, a large number of innovations should result as the medical community strives to make US healthcare more cost efficient.

To take advantage of these opportunities, hospitals and other medical facilities will need to make investments in equipment, tags, and software, along with an understanding of the value of unique identification in everyday operations. Concern exists within the medical community that because of pressure for hospitals [20] to have the most advanced equipment for the treatment of illness, there will be little surplus money available for RFID investments. One administrator has commented, "'Hospitals aren't ready to adopt RFID from a financial perspective and because of the hospital IT infrastructure. They're barely ready for bar-coding…I see in 10 to 20 years bar-coding still being used at hospitals, as opposed to RFID." [21]

This is a valid criticism given the cost of RFID technology and the complexity of medical operations. However, the alternative is not appealing either. Foregoing investment in a key technology such as RFID will likely make medical operations even more complex in the future, as a lack of information inhibits treatment. Given that the cost for RFID technology should decrease during the next few years, the benefits of the technology will become apparent. In preparation, hospitals need a framework in place to evaluate the costs, clinical/operational benefits, reliability, the path to technological implementation, along with the issues surrounding liability. [22]

The practice of medicine is a complex matter involving a number of different constituencies. Doctors, healthcare workers, and administrators must take a broad view concerning the EPCglobal Network and RFID technology because the principles of unique identification affect so many different aspects of medicine.

One area that has gained a great deal of attention involves medical records. Overwhelming support, 75% of those polled, exists for moving to a system of electronic medical records that would include patient information and digital images. [23] However even though such a system could "trim 20% off the nation's $1.6 trillion healthcare bill, reduce the alarming number of medical mistakes, and improve the quality of the nation's health care" [24] the conversion costs are staggering. Some estimates put the cost of a national, computerized health care record system at "more than $300 billion over a decade."[25]

Many organizations and companies have attempted to standardize medical records in the past, with little success. Part of the reason may be the complexity and variety of the data. Another may be the individuality and variability of clinical care by both the institutions and physicians. Still another reason may be that standardization is typically implemented and recognized within professional medical specialties and not across the entire healthcare space.[26] Finally, there is the complex issue of privacy and security of medical histories.

Whatever the reason, the net result is that there is no simple, efficient and agreed upon method to share medical information. This becomes more than obvious during any visit to the clinic or trip to the emergency room. Basic health and status information is gathered at every point, and each format is different.

As a fist step in solving this problem, the EPCglobal Network and RFID technology provides a standard way to describe physical objects. With a common identifier, relevant data can be linked, thus improving information handling efficiency. While much more work needs to be done in the area to formulate a comprehensive solution, having standards for unique identification is a good start.

CHAPTER 9

Agriculture: Animal Tracking

The publication of *The Jungle* in 1906 by Upton Sinclair focused national attention on the sanitary conditions of the meat industry and the potential for extensive outbreaks of disease from the consumption of tainted food. Given a wave of public outrage, US Congress passed the Meat Inspection Act along with the Federal Food, Drug, and Cosmetic Act during the same year that Sinclair's book first appeared in print.[1] This represented a major step forward in ensuring the safety of both fresh and process foods.

Though food technology has advanced a great deal since 1906, there continues to be concern about the safety of fresh foods that are the unprocessed product of farms. Prone to contamination from salmonella, E. Coli, and other bacteria, the freshness of high protein foods such as meat, fish, poultry, and eggs is especially important in controlling outbreaks of foodborne illnesses.[2] By one account, there are 76 million illnesses and 5,000 deaths each year from food contaminated with various pathogens.[3]

The appearance of Bovine Spongiform Encephalopathy (BSE) or "Mad Cow Disease" in the U.K., Canada, and the United States along with other parts of the world has raised additional questions about food safety because the disease has a long latency period that spans several years. In this case, manifestation in humans would be difficult to associate with specific sources of foods, causing great uncertainty in linking cause to effect.

Because of all of these factors, there has been a growing interest in implementing tracing systems within agriculture, specifically the livestock industry.[4] The basis of any system of this type depends on some form of unique identification. In many ways, establishing a tracing system in agriculture bears similarities to the pharmaceutical and medical industries as discussed in chapters 7, and 8. However, there are basic differences that make agriculture a unique situation regarding the application of the RFID technology.

For example, all agricultural supply chains share a commodity orientation where production of like goods takes place on numerous farms. The complexity of the agricultural supply chain arises because agricultural products often have significant variations in taste, vitamin, mineral, and protein content, bacterial contamination, and numerous other attributes that define the quality, safety, and identity of food. These variations sometimes depend on the location where production occurs, but it is also true that variation is a function of methods employed in agriculture, which do not always follow a standard practice.

Along with biological variation, there is also the inherent characteristic of raw material mixing. The output from farms must undergo various forms of processing before eventual consumption. This makes the task of maintaining unique identification throughout the supply chain difficult.

In the agribusiness environment, food traceability serves several different functions for various constituencies including identification of the origin of contaminated food (public safety), the limitation of liability in the event of disease outbreak (business), and information about inferred physical quality characteristics (consumer).[5] Within the US livestock industry, most of the emphasis concerning traceability relates to safety and liability issues. Consumers tend to depend on branding for making quality and purchase decisions rather than a comprehensive understanding of the origin of specific food products.[6]

In addition to the underlying complexity of agricultural supply chains, there is also the established trend of globalization that adds new dimensions to traceability. Though there are many benefits from world trade, the crossing of borders increases the risks that various types of contamination might spread quickly worldwide.[7] Trading partners from around the world are increasingly interested in establishing agricultural tracing systems as a means of mitigating the negative economic consequences of disease outbreaks.

With these characteristics in mind, this chapter introduces the issues surrounding traceability in agriculture. Although few public or private organizations have announced specific plans for implementing the EPCglobal Network within agriculture, there are several important examples where RFID technology will play a role. The next section takes a deep look into the US livestock industry, where traceability has surfaced as a major safety and business issue.

Disease Threat and Animal Tracking

The largest single segment of the US agricultural economy, beef sales are more that $50 billion per year at the retail level.[8] The scale of the industry is massive. Figure 9-1 shows the various steps of the beef supply chain, from production through sale to consumers.[9]

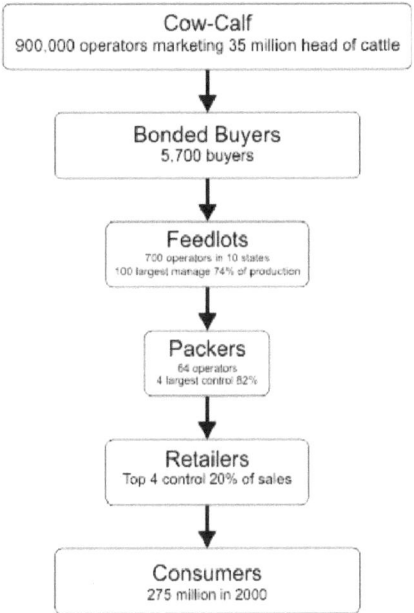

Figure 9-1 – The Beef Supply Chain

Like all commodities, beef is subject to price swings based on supply and demand. Factors such as the price of feed, time lags in herd growth, the existence of a futures market, and the prices of substitutes such as pork and chicken all play an important role in the beef commodity cycle, which had followed a predictable trend for many years until 2003.[10]

During 2003, the US Agricultural industry was at the peak of the commodity cycle in almost all areas. Strong demand from overseas for American soybeans and meat pushed commodity prices up. A consumer shift toward high protein diets stimulated domestic demand for beef, pork, and chicken, and the gradual increase in petroleum prices meant more demand for ethanol, which diverted corn from traditional food processing and feed uses of lesser profit margin. The farm economy appeared to be recovering from a multi-year recession characterized by low commodity prices and financial hardships for many Midwestern rural communities. For the first time in five years, farmers began to consider purchases of new equipment and investments in their operations.[11]

Then in December 2003, a single case of Mad Cow disease was discovered in Washington State.[12] Almost overnight, more than 40 countries slapped a ban on beef imports from the US. On an annualized basis, beef exports equal $3 billion in revenue and about 10% of the domestic beef market.[13] Virtually all of the US export market was lost.

In the next few months, prices for live cattle decreased about 20% and several meat packers began to layoff workers. Some regional economists projected that 50,000 jobs in the Great Plains states could be lost if the ban continued.[14] Fears existed among livestock producers that major trading partners such as Japan and Mexico would seek alternative sources of beef from places such as Australia, New Zealand, and South America, causing long-term structural damage to the US beef industry.

Previously, Canada had reported a case of Mad Cow disease seven months earlier (May 2003) resulting in a near worldwide ban on imports to other countries. This was a particularly bad development because at the time 50% of Canadian beef was sold to other countries. Prices dropped by about half and many ranchers went out of business. The overall Canadian farm economy recorded a net loss for 2003 in spite of heavy subsidies from the government to prop up the industry.[15]

Since cattle freely move across the border between the US and Canada as part of a complex supply chain that includes trade in live dairy and beef animals, it was just a matter of time before discovery of a case of Mad Cow disease in the US. During the 1980s, Canada had imported animals from the UK where Mad Cow disease was identified in 1985.[16] Because of the introduction of contaminated meat into the supply chain processed from animals with Mad Cow disease, the UK had experienced over 100 deaths during the past ten years.[17] What is particularly alarming about this disease is that the incubation period lasts from 10 – 16 years. Luckily, the UK took many steps during the 1980s to limit the spread of the disease or it is possible many more people could have died.

Upon discovery of Mad Cow disease in Washington State, US Department of Agriculture (USDA) scrambled to find information about the extent of the problem. With no national system of cattle identification or tracking, the USDA sifted through paper records kept at local farms for information on the details surrounding the movements of the infected cow. The process required several weeks before an understanding of the magnitude of the problem became clear.

Initially, government investigators believed the infected Washington cow was among 80 that were raised on the same farm in Canada and eventually sold to various interests in the US. Since contaminated feed was probably the cause of the outbreak, USDA officials felt it was reasonable to conclude that the other cows in the herd were also exposed to Mad Cow disease and perhaps infected. At the conclusion of the investigation in February 2004, only 27 of the 80 were positively identified.[18] Likely, the other cattle in the group had already entered the US food supply by the time the first case appeared. The USDA stated that it was impossible to determine which retailers might have sold the meat. In retrospect, a lack of information prevented development of timely answers to basic questions about the infected cow including; "Where did this animal come from, where were the feed sources, where did animals move out of this herd?"[19]

The reality of modern agriculture is that feedlots, dairy farms, and other animal growing operations are potential breeding grounds for a number of diseases that could threaten the US food supply. The close proximity of animals in these operations means that rapid transmission of bacterial and viral diseases is highly probable and use of animal remains as a protein supplement for feed, a practice banned in the US in 1997 for beef feedlots, can lead to unanticipated outcomes such as the spread of Mad Cow disease. Dealing with these situations means that the USDA must have the best information to identify outbreaks and take action needed to reduce the spread of animal diseases.

In the aftermath of the US Mad Cow disease outbreak, several systems have been proposed to improve animal identification and tracing capabilities within agricultural supply chains. Most notably, a company named Digital Angels has technology that allows veterinarians to implant an RFID chip into animals as a means of tracing and monitoring.[20] The technology allows a "biography" to be assembled for each animal that would include origin, location and health status of the animal as it moves through the supply chain. In addition, the company has a new microchip that measures the body temperature of the animal. Since elevated body temperature is a reliable indication of disease, a quicker diagnosis might result through constant monitoring.

In another development, Swift & Company announced a new system that provides the capability to trace boxed beef back through the supply chain to the feedlot.[21] Eventually, this system will feature unique identification of each animal through technology developed by Optibrand, Inc. that scans an animal's retina and possibly an RFID tag placed on the ear as a means of unique identification, and then records information into a database. The scanning process will also use global positioning system information to provide location data for any animal, at any farm supplying Swift with live cattle for slaughter.

Though both of these technologies provide a means of tracing within the livestock supply chain, there are issues concerning cost, the use of proprietary systems, and interoperability. Given an outbreak of disease that might involve numerous farms, meat packers, and retailers, the USDA has maintained the need for a national system of identification and tracing that spans all entities within the livestock supply chain. Such a system would allow authorities to "pinpoint a single animal among the nation's nine billion cattle, pigs and chickens." [22]

Setting a mandatory participation date of January 2009, cattle would be identified and monitored using a combination of technologies including RFID ear tags, retinal scans, and DNA testing. Other farm animals such as hogs and chickens generally move in groups through the supply chain and the USDA has proposed different methods for traceability. As of early 2006, the USDA was holding public hearings on the implementation of a national animal ID system.

Several private groups have proposed alternative national tracking systems that they say will achieve the same goals at lower cost.[23] In one case, a Montana livestock auction company named Northern Livestock Video recommends a system called Verified Electronically ID Source and Age Program (VESA). [24] Acting much like product authentication in the pharmaceutical system, the approach verifies origin and age of animals without constant monitoring as they are transferred within the supply chain. The system works as follows:[25]

1. The producer would purchase RFID tags from their state veterinarian.
2. When a calf is born, the producer would send the animal's birth data along with address information to the state.
3. Upon receipt, the state would enter the information into a national database. This data is then encoded into the RFID tag sent by the producer.
4. The state would return the tag to the producer, who would have a veterinarian affix in the animal's ear as part of the health inspection process prior to a sale.

The advantage of this approach is relatively little investment in computer equipment to monitor RFID tags. No readers or computers would be required at the farm level. This would allay concerns by producers about the amount of data generated from constant monitoring as required by the USDA plan. In the words of one stockyard operator who turns over 450,000 head of cattle each year, "The wad of data that would go with that would be almost impossible. God Himself couldn't do it."

Beyond the logistical aspects of managing RFID data as part of a tracing system for livestock, there are also technical limitations that need further research and development effort. Most of the existing passive tags used for animals operate at low frequency, which means the reader must be within about ten centimeters to retrieve data and the transmission rate is slower. Though these types of RFID tags can be implanted under the skin of pets, there are limitations when applying the same tag to farm animals. In this situation, it is nearly impossible to get close enough to the reader for a successful scan especially when some farm animals weight more than 500 kilograms and tend to thrash about when in confined areas such as narrow corrals or stanchions.

One group of researchers has proposed the use of passive UHF tags that operate in the 915 MHz range as a better approach.[26] These tags could be affixed to the ear of the animal. With read ranges in excess of five meters and faster data transmission rates, this allows for greater practicality in scanning animals as they move through stockyards.

However, one significant limitation with this approach exists. With passive UHF tags, the size of the antenna is much larger as compared to low frequency tags. The configuration of the antenna is such that it might be too large to fit on an animal's ear. Research continues to focus on alternative antenna designs for general use in livestock agriculture.

With these advances in identification, meat processors and the USDA will eventually have complete tracing capability within the entire beef supply chain. In the event of a future problem, such as another outbreak of Mad Cow or other highly contagious animal illness like hoof and mouth disease, the improved track and trace technologies will enable sense and respond capabilities that might ultimately save lives. This type of capability ensures the safety of the food, and strengthens the agricultural industry in the face of potential bans on US beef imports from other countries. World customers for US agricultural products increasingly desire more information about the production and handling of such products as meat and grains. Following the announcements by Wal-Mart, the Department of Defense, and the FDA, concerning implementation of the EPCglobal Network and improved track and trace technology, it is likely that further developments concerning RFID in the agricultural industry will occur in the future.

CHAPTER 10

Food: Dynamic Expiration Dates

Many of the RFID applications in agriculture have focused on developing a tracing capability in the event of an outbreak of disease. While this is an important part of ensuring food safety, there are other potential applications of RFID technology that relate to the quality of processed food. In this regard, the EPCglobal Network and RFID enable emerging sensing technologies that can monitor the essential factors responsible for affecting the flavor of food after packaging.

In the field of food science and technology, there are a number of different processes to achieve food preservation on a large scale. Perhaps one of the oldest, the idea of using heat to destroy the bacteria, molds, and yeasts that cause spoilage and foodbourne illness, dates to 1809 when Nicholaus Appert won a cash prize given by Napoleon Bonaparte for preserving food contained in large, sealed glass bottles that were heated in boiling water.[1] Napoleon became interested in food preservation as a means of provisioning his troops while they were on the march during foreign military campaigns.

Many years later, Louis Pasteur discovered in 1862 that controlled heating at a predetermined temperature for a fixed amount of time destroyed specific microorganisms present in fluid foods.[2] The discovery that heat destroyed pathogenic bacteria has protected countless numbers of people from exposure to various diseases associated with fresh milk such as tuberculosis.

The work of Pasteur also had an important impact on others. After reading about the pasteurization of milk, Dr. Thomas Bramwell Welch of Vineland, N.J. applied similar heating techniques to 40 gallons of freshly squeezed Concord grape juice in 1869.[3] He observed that the juice, contained in sealed glass containers much like what Appert had done, remained unfermented after treatment with heat. The use of five-gallon glass containers as a commercial means for the preservation of fruit juice continued for many years until the development of large, steel, storage tanks during the 1920s. This single event, the pasteurization and storage of Concord grape juice by Dr. Welch, was the birth of the processed fruit juice industry in the US, which now equals billions of dollars in sales each year.

Though heating is effective in achieving food preservation, it also has inherent downsides that can have a significant impact on consumer acceptance. Heating destroys some of the delicate organic chemicals responsible for flavor in food.[4] In the case of fruit juice, naturally occurring chemicals called esters are highly sensitive to heat. Further, preserved foods in glass or plastic bottles, metal cans, or flexible pouches can experience additional flavor loss after processing because chemical reactions continue to take place.

The most common post-processing chemical reaction is oxidation, which is dependent on the levels of oxygen inside the package along with temperature and other factors such as motion.[5] Food processors control the rate of oxidation through packaging designs that limit the amount of oxygen in contact with food along with controlling storage conditions. However, in spite of a number of techniques designed to reduce the speed of chemical reactions, it is impossible to totally stop flavor loss in preserved food products. As a general rule of thumb, the speed of chemical reactions responsible for flavor loss doubles with each 10 degree Celsius increase in temperature.[6]

To deal with this reality, food processors establish a standard shelf life for almost all products sold in retail stores or food service establishments. This provides a rough indicator of time-dependent quality for the benefit of upstream links in the supply chain such as retailers, distributors, and consumers. The method to establish shelf life usually depends on test data from keeping quality studies (KQS), where technologists expose a specific food product to elevated temperature and then evaluate flavor at various intervals of time.

Based on the results of the KQS, food and beverage manufacturers establish a shelf life for a product at the time of production. Most often, a "use by" date appears on the package, sometimes as a code, to alert consumers and others within the supply chain that the food might be beyond acceptable flavor.[7] This printed expiration date remains fixed during the life cycle of the product.

The lack of a "dynamic" shelf life indicator that takes into account storage conditions and other factors responsible for diminishing flavor has practical economic consequences. For example, beverages such as beer begin losing flavor almost the moment after bottling.[8] If a consumer purchases an old bottle of beer, and it tastes stale or "skunky," the chance of repeat purchases is reduced.[9] This is an extremely important consideration because the business model for a food processing firm is built upon the assumption of repeat sales.

To achieve a dynamic shelf life indicator that takes into account environmental conditions such as temperature requires a way of obtaining real-time data on the state of a food product as it travels through the supply chain. Given this information, various types of mathematical models can calculate the time remaining for optimal freshness and communicate this knowledge to consumers along with those responsible for managing inventories. This predictive modeling approach holds much promise in reducing the chances that consumers will unknowingly purchase a food product that has unacceptable flavor. In addition, predictive models can also play a role in reducing spoilage for fresh products such as fruits and vegetables.

The next section takes a deeper look into the development of dynamic expiration dates and predictive modeling using the EPCglobal network, RFID technology, wireless communication, and sensors. While the test application put forth deals with food used in the military, the underlying technology has applications in civilian supply chains for anything that is temperature sensitive.

Tracking Flavor Electronically

Perhaps one of the most important aspects of an Army involves the provisions required for the troops in the field. Beyond the logistical issues associated with maintaining adequate lines of supply for critical items like ammunition and medical gear, the variety and flavor of food plays an important role concerning the health and morale of a fighting force.

The unique properties of military operations require that food for the troops must meet a number of criteria that have no equivalent in civilian supply chains. Besides being of high nutritional value to enhance troop performance under intense periods of physical activity that might include battle, the food supply must have a reasonable cost along with self-heating capabilities for field operations.[10] Given that transportation is always of great importance in maneuver, military food must occupy the minimum cube and have the lightest possible weight. This is particularly important if the food is carried into battle by the troops. Finally, as with all military supplies, the food must be able to withstand rough handling that might include airdrops via parachute.

Since military conflicts seldom take place under ideal circumstances, significant extremes often occur in terms of temperature. For example, severe heat such as experienced in Iraq can reduce the shelf life of the military's Meals Ready to Eat (MRE) from "36 months to just one month." [11] Though the use of time and temperature indicators (TTI) [12] on each case of MREs has provided greater visibility concerning expired product, these label-based indicators require manual reading and there is no mechanism to predict when product quality might become unacceptable. Figure 10-1 shows a picture of the contents of an MRE.[13]

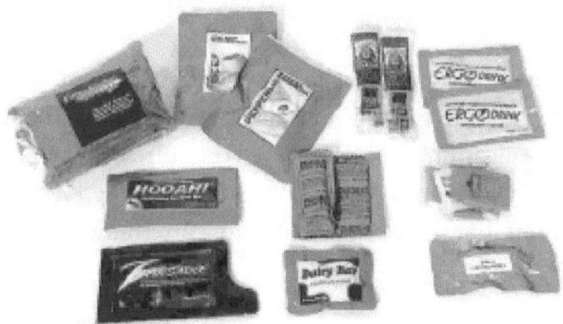

Figure 10-1 – The MRE

Used with permission, US Army Soldier Systems Center – Natick

A System for Dynamic Expiration Dates

Unique identification forms the future basis for a number of automated functions within the supply chain. Nowhere is this more apparent than in a US Army prototype of a system that calculates dynamic expiration dates for MREs.[14]

Using temperature sensors attached to pallets of MREs allows for real time information on the state of a specific product that can in turn become the inputs for mathematical models designed to update the predicted shelf life of the food.[15] With this approach involving a combination of the EPC, sensors, and mathematical models, the calculation of a dynamic expiration date becomes possible in addition to a historical record of heat exposure.

The prototype tested by the Army included battery powered, semi-passive sensors designed to report real-time data on temperature.[16] These temperature sensors transmit data at 2.4 GHz.[17, 18] Each box of MREs also contained a standard, 915 MHz RFID tag of the type used in the consumer goods industry. Figure 10-2 shows a picture of a temperature sensor.[19]

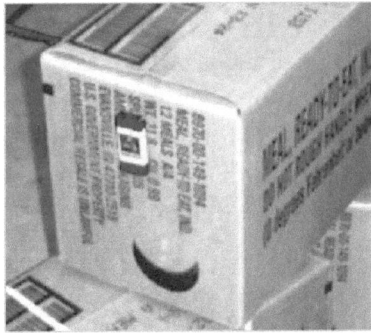

Figure 10-2 – A Temperature Sensor

Used with permission, US Army Soldier Systems Center – Natick

Even though a box of MREs contains a number of different types of foods, breakfast, lunch, and dinner, real-time information concerning the storage temperature of the entire unit is still useful in determining overall shelf life. For the prototype, a single sensor continuously transmitted temperature data and the EPC for each pallet, which was gathered by readers within close proximity. The EPC contained on the passive tag for each box of MREs was associated with the pallet identification EPC through the middleware. When a reader identified an EPC for a particular case, it was linked to the temperature record for the pallet from which it came.

Given this type of system, sensing capabilities to monitor environmental variables are not limited to just temperature. There are a number of opportunities to link other types of sensors to active RFID tags. For example, sensors can measure variables such as light, humidity, pressure, vibration, sound, acceleration, current draw, motion, among others.[20] In almost all cases, the data stream is not the actual measurement taken at some time interval, but rather the deviations as compared to the pre-existing state of the measurement.[21] With this approach to measurement there are sometimes concerns relating to calibration, time lags, and drift because small errors in the state of initial conditions tend to amplify through time. Future research in this area is expected to improve the accuracy of sensors.

Extensions to the Civilian Supply Chain

While the use of sensors in the fresh food supply chain is not yet common, there are a number of applications under consideration that have potential to reduce the waste resulting from spoilage. For many fresh fruits and vegetables, the shelf life is as little as one week.[22] This means that any spike in temperature or time lag in transportation can become critical to freshness. In addition, the consumer trend toward purchasing prepared meals and other items such as sushi at the retailer means temperature control becomes an important factor to reduce spoilage and the incidence of food poisoning. Although temperature is perhaps the most significant factor to monitor, specific foods also require other types of monitoring that are no less important in controlling quality.

For example, high value fresh fruits such as strawberries are particularly sensitive to bruising during transport. This has significant economic consequences because about 10 percent of all perishable goods become unusable before reaching consumers.[23]

In the case of strawberries, it is possible to attach vibration sensors to shipping containers allowing for measurement of the number of bumps encountered during over-the-road transport.[24] With this information, retailers can analyze the cause and effect between poor roads, distance of transport, and the quality of delicate fresh fruits like strawberries. These types of measurements are also significant for other purposes such as understanding the stresses encountered by food packaging. In the situation of glass packaging containing fluid foods, hydrodynamic stress caused by road vibration can be a significant factor in breakage.

Beyond sensing, RFID can have other favorable impacts on shelf life through tracking and tracing capabilities. Tanimura & Antle, a produce packer from Salinas, California, uses the tracing capabilities of RFID as an integral part of logistical decision-making.[25] When farm workers harvest heads of lettuce, the product is wrapped and placed in a re-usable plastic container that contains an RFID tag. The EPC helps to identify where the lettuce was grown and the time that the lettuce was picked.

Given that harvest operations take place over a wide area, a number of trucks converge at the central refrigerated warehouse nearly simultaneously. Since chilling of the lettuce is the first step in packing, a bottleneck sometimes develops. When the tagged containers of lettuce arrive at the refrigerated warehouse, each is scanned and the EPC information recorded in a database. This information is vital in helping warehouse managers prioritize which containers to place into refrigerated storage first. By giving top priority to refrigerating the lettuce with the longest time since picking, the shelf life of this product can be extended. In the case of lettuce, getting the product chilled even one hour quicker can make a significant difference in extending shelf life.

Though uncertainty exists concerning the ROI associated with using sensor networks for monitoring processed and fresh foods, there is optimism that value will result from reducing spoilage losses and monitoring the freshness of processed foods. A major factor in achieving a return will be the declining cost of sensors to detect temperature. The current prices of up to $20 per unit preclude widespread use within the food industry.[26] However, as different applications develop, some expect prices to drop to about $1 per sensor.[27] At this cost, widespread use in conjunction with the EPC is possible. Continuing research at academic institutions such as the University of Florida should refine the application of the EPC along with sensor networks in the fresh fruit and vegetable industry.

CHAPTER 11

Retailing: Theft Prevention

Based on his years of experience in retailing, Sam Walton, the founder and former CEO of Wal-Mart, once commented that theft is "one of the biggest enemies of profitability in the retail business."[1] Unfortunately, his observation continues to remain valid today. According to various surveys, theft is a serious problem within supply chains costing North American retailers and manufacturers at least $25.0 billion each year. [2]

While theft has historically taken forms such as shoplifting from stores, new technologies give thieves the capability to attempt increasingly sophisticated forms of theft. In 2003 a web site named Re-code.com offered Internet users barcodes that could be printed at home on stickers, taken to a store such as Wal-Mart, and applied over the top of bar codes already existing on products.[3] The web site encouraged visitors to "name their own price" as a political protest against the pricing policies of large corporations.[4]

For example, a customer could change in-store prices by "sticking a bar code for generic cereal on a name-brand box in the store."[5] By using the automated checkout lanes, which are not closely monitored, customers would receive the lower price. This type of theft is extremely hard to control and exploits the weaknesses of bar codes along with advancements in home printing capabilities. Wal-Mart took quick legal action to shut the site down.

Though theft has been an ongoing problem, it has received comparatively little attention in business because of the difficulty in finding a solution capable of withstanding the test of time and the almost limitless ingenuity of thieves. As a result, losses from theft have remained constant at about 1.1% of sales with no improvement during the past ten years.[6] The ongoing cost of theft directly reduces net income dollar for dollar. For a stolen item with a profit margin of 10%, revenues must increase by 10 times the amount of the theft to recover the net income lost.[7]

With a general lack of pricing power in most consumer goods markets, few firms are in a position to cover these losses through higher sales volume or increased prices. Industry needs a comprehensive solution to this continuing problem.

Theft is part of the broader category of shrinkage and is hard to pinpoint with accuracy. Total shrinkage, as measured by inventory adjustments, is the only true indicator of theft; however, this category also includes other losses from process failures, spoilage, accounting errors, and vendor fraud.[8] Few, if any firms know precisely the amount of theft that occurs each year from their stores, manufacturing plants, and warehouses. The consensus estimate puts theft at 75% of total shrinkage.[9]

Complicating matters, current investments in technology to reduce theft in one area of the supply chain frequently achieve mixed results. Theft seldom totally disappears. It tends to shift, appearing in other parts of the supply chain where security measures are soft. A truly effective theft containment solution must address losses across the entire supply chain.

This chapter analyzes both internal and external theft within the supply chain along with technological methods for controlling theft. The discussion concentrates on the implementation of RFID technology as a means of predicting, preventing, detecting, and providing proof of theft for high value items that are compact in size, easy to conceal, not immediately perishable, and with high resale value.

Theft is not just a problem experienced in retail stores. Financial loss from cargo theft totals at least $10 billion per year for the United States alone.[10] This is a staggering amount for businesses to withstand.[11]

Besides direct monetary loss, theft affects firms in other more subtle ways that often add indirect costs, reduce customer service, and limit revenues. The next section outlines a few of these hidden impacts to retail and manufacturing operations.

The Indirect Impact of Theft

When theft occurs, computerized perpetual inventory systems get out of synchronization with physical counts. Inventory records become inflated because there is no accounting transaction for goods stolen from manufacturing plants, warehouses, or stores. This is particularly important because many Enterprise Resource Planning (ERP) systems use a perpetual inventory method called "back flushing." With this technique, inventories are adjusted for depletions based on a paper transaction with ongoing re-calculation of inventory balances. Comprehensive physical counts occur only once or twice per year to verify the calculated inventory level.

Given this type of system, theft causes inaccurate inventory records that may be in error for many months. The cumulative effects are devastating to customer service because many stores and warehouses depend on continuous replenishment systems, and accurate inventories, to maintain stock levels for thousands of items.[12] Out-of-stocks are a particular concern to retailers. According to a recent study, nearly 23% of consumers leave a store immediately in response to an out-of-stock.[13] No retailer can afford this loss of business.

Pushing Responsibility Upstream

Increasingly, retailers are requesting development and manufacture of "theft proof" items. Often this involves expensive packaging modifications such as altering the size of small high-value items to make stealing more difficult. Many retailers also ask that manufacturers make investments in store fixtures such as locked cabinets in an effort to reduce theft.

This is consistent with the gradual trend in industry toward supply chain integration. Retailers are shifting more responsibility for many aspects of store operations upstream to the suppliers, including such practices as vender managed inventory and category management. Theft reduction follows this trend.

For example, manufacturers are often required to attach electronic article surveillance (EAS) tags to products and packaging during production (this is called source tagging). Each retailer currently uses one or more of several proprietary systems for the tags. Since no single, universal standard for EAS tags exists, source tagging results in an exponential increase in SKU's for manufacturers adding considerable complexity, and cost, in managing inventory.

Finally, a recent trend exists toward consignment sales (also called scan based sales) where manufacturers receive payment from retailers upon scanning at checkout and are responsible for customer service and shrinkage within the supply chain up to the point of sale. In this situation, the manufacturer will absorb all retail theft losses occurring in warehouses or retail stores prior to actual purchase by consumers. With no integrated program for theft control, implementing a consignment sales program represents an immediate and substantial negative liability for the profit margins of manufacturers.

For all these reasons, understanding and controlling theft across the entire supply chain is important for manufacturers and retailers alike.

Defensive Merchandising

Besides operational impacts on inventory accuracy and consignment sales, theft also causes indirect constraints to merchandising. Retailers often limit the number of items on display to control theft. This practice is termed "defensive" merchandising. Some retailers go farther and do "restrictive merchandising" by placing goods behind counters or using locked display booths. In these cases, retailers will leave a dummy package or sign on the shelf directing customers to the counter for eventual purchase.

While these two policies prove effective in reducing shelf stock loss from theft, each method also limits sales because customers must request assistance to make a purchase. Marketing executives at one firm estimate that sales would increase up to 75% by removing restrictive access and defensive merchandising.[14] In this case, the estimate has a firm analytical basis in market research conducted in the United States and Europe.

Though defensive and restrictive merchandising is a common tactic used in retailing, reported sales increases from removing these constraints depend on a number of variables. Firms should not interpret anticipated sales increases broadly, although it is logical to expect that removing merchandising constraints should improve sales to varying degrees.

Theft is not just an issue of direct financial loss but often the tactics involved in controlling theft place burdensome limitations on business operations, which often decrease revenues. Retailers cannot ignore the potentially huge positive impacts of removing specific constraints such as restrictive access or defensive merchandising, or improvement in inventory accuracy.

Analyzing Theft

Given the size of the theft problem in North America, several studies have attempted to analyze the reasons why people steal.[15, 16, 17, 18] These studies assume that understanding the psychology of theft, pre-meditated behavior or impulse, will lead to better methods of prevention. However, it is difficult to develop a classification of thieves that has meaningful use in practice.[19, 20] Some argue that a precise classification scheme is not the right direction.[21] Rather, emphasis should focus on the mechanics of theft and technological solutions that will make theft hard to execute regardless of the class of thief or personal motivations. Typing thieves into psychological constructs as a means of targeting certain types of theft, misses the emerging consensus among criminologists that thieves do not fit pre-determined patterns.[22]

An effective solution to theft will ultimately depend on both technological and psychological deterrence. Theft of merchandise is a "victimless" crime in the minds of thieves.[23, 24] For many years, the simple use of employees as "floor walkers" was enough to deter some types of theft. The floor walkers made crime seem less of a victimless offense by building relationships with customers.

Many retailers continue to employ greeters and store staff to approach customers in a positive way and to serve as a watchdog for suspicious behavior in the store. The presence of store employees on the floor, though underutilized as a deterrent, is an effective means of reducing theft.[24]

Some thieves actively calculate the risk of apprehension. In this case, methods of detection based on rigorous technology play an important role in producing uncertainty in the mind of a thief. Deterrence through the rational expectation of penalty makes the odds of a successful theft look less attractive. However, a technology that relies on stimulating fear of apprehension during theft will not be effective if the thief disregards or is unaffected by the stimuli generated by technology.

Characteristics of RFID Technology

In developing an approach to combat theft, it is important to highlight a few characteristics of RFID technology that provide unique anti-theft capabilities.

No Line of Sight Identification

One of the primary advantages of Radio Frequency Identification (RFID) technology is that it does not require a direct line of sight between the reader and the tag. The RFID reader is capable of communicating with a tag through optically inert mediums such as the package containing the product or even the product itself. This is in contrast to barcodes that require unobstructed alignment with a reader.

Mass Serialization

The Electronic Product Code (EPC) permits the assignment of unique serial numbers to identify discrete manufactured objects and object aggregates (e.g. secondary packaging, pallets etc.). The primary goal of this numbering system is to ensure that any manufactured object can have a unique identifier. The identifier is very important. It not only provides ubiquitous unique identity for objects but it is also a key to access data on a network about the object.

Real-Time Visibility

EPC Tags along with the RFID information infrastructure provide the capability to monitor objects moving through the supply chain in real time. This enables companies to account for inventory at all locations. Because of open standards, RFID is designed to integrate with other technologies, such as closed circuit television (CCTV) and motion detectors, to provide security staff with information as crimes transpire.

Track

Together, RFID capabilities such as mass serialization and real-time visibility allow detailed tracking within the entire supply chain. Tracking is defined as maintaining control on a particular object going forward in time. An example of maintaining control is the process of pre-positioning information during the execution of a transaction like the Advanced Shipment Notice (ASN). The EPCs of products that comprise a shipment can be sent in advance to the receiving company. When the goods arrive, the EPCs are checked off the advance list. Missing, different, or excess EPCs are then investigated.

Trace

Besides tracking, RFID also provides unique capability to do tracing within the entire supply chain. Traceability, used interchangeably with the term "pedigree," is the ability to build the supply chain history for a uniquely identified object, i.e. location, ownership, telemetry etc. This may involve accessing data in distributed databases across a number of companies. The PML Service/Server is an important component of this tracing capability.

Table 11-1 summarizes the various characteristics of RFID that pertain to anti-theft applications and the components of RFID that enable these characteristics:

Table 11-1 – Characteristics of RFID Technology That Relate to Theft

	Tag Reader	EPC	ONS	Middleware	PML	PML Server
No Line of Sight Identification	X					
Mass Serialization		X				
Real-Time Visibility	X			X		
Track		X	X	X	X	X
Trace		X	X			X

A Conceptual Model for the Analysis of Theft

The advantages of RFID technology opens many possibilities for anti-theft applications in retail stores and manufacturing firms. However, before exploring specific RFID applications, an initial conceptual model of the process of theft must exist to guide thinking about the appropriate use of the technology. Based on work initially conducted at the MIT Auto-ID Center, the next sections discuss three basic steps that form a simple conceptual model of theft.[25]

Before Theft

Theft is a complex process that takes many different forms. It is extremely difficult to pinpoint the exact time when a thief makes the decision to steal but has not yet executed the theft. In some cases, the decision to steal is conscious and rational, made before entering a store, warehouse, or manufacturing facility. In other cases, theft occurs as an impulse while the thief is close to merchandise. Because of these reasons, it becomes difficult to establish a uniform decision-making process for theft. However, it is clear that certain types of behavior precede different types of theft. If detection of these behaviors can take place then the opportunity exists to design systems to deter theft before it ever occurs.

For a number of safety and legal reasons, it is best to detect in advance and deter theft as opposed to dealing with theft after it has occurred. Detection of a potential theft can occur through a combination of human intervention, such as store personal on the floor, in addition to various detection and monitoring technologies, like closed circuit televisions (CCTVs) and RFID technology, which might give information about suspicious behavior that often precedes a theft. RFID provides unique predictive potential through the ability to monitor the sudden, simultaneous disappearance of a number of tags from the reader field and the ability to flag unusual patterns of product movement.

RFID is most effective as a theft predictive and deterrence measure when combined with other antitheft sensing devices. The various components of the technology also provide a dynamic platform to integrate different measures over time.

For example, the technological ability to predict theft allows the option of enabling deterrence measures. This might be in the form of a Light Emitting Diode (LED) display that acknowledges the person and the quantity they have removed from the store shelf. This can act as reinforcement to posted signs directed at potential thieves that warn of theft monitoring activities within a store. A combination of all of these measures is enough to discourage some thieves from attempting a theft.

During Theft

Even the best theft deterrence system will not work for some thieves. When a thief has physically taken merchandise (concealed or unconcealed) with the goal of exiting a store, warehouse, or manufacturing facility, a potentially dangerous situation exists. Detection of a theft in progress can be accomplished through a number of different technologies such as EAS, CCTV, and RFID. In this regard, RFID has great potential to detect concealed merchandise.

However, if successful detection does take place, there is always the question of what to do to stop a theft in progress. In many cases, employees are instructed by management not to confront a thief attempting to exit a facility because the chance of physical injury to the employee is not worth the value of the recovered merchandise.[26] Though a theft in progress can be detected, it is unclear if this information has any practical value in preventing theft at the time it occurs.

After Theft

Even with comprehensive security measures, there are situations where thieves will succeed in stealing items. At this stage, RFID, the EPC, and trace capabilities can help law enforcement authorities to prove the pedigree of a particular item that might be sold in secondary or black markets. This helps in proving if an item was legally purchased, including status – returned, recalled etc. The ability to determine the status of a particular item could potentially lead to the disruption of the goods to cash conversion cycle for stolen merchandise. Over time, this should reduce the incentive for professional thieves.

These stages of theft, before, during, and after, form a basis for analyzing the application of RFID technology as a means of theft reduction. As mentioned previously, theft is a complex human process that is hard to characterize with consistency. Figure 11-1 is a flow diagram intended to visualize the steps involved in conducting a theft, though at best it can only serve as an approximation.

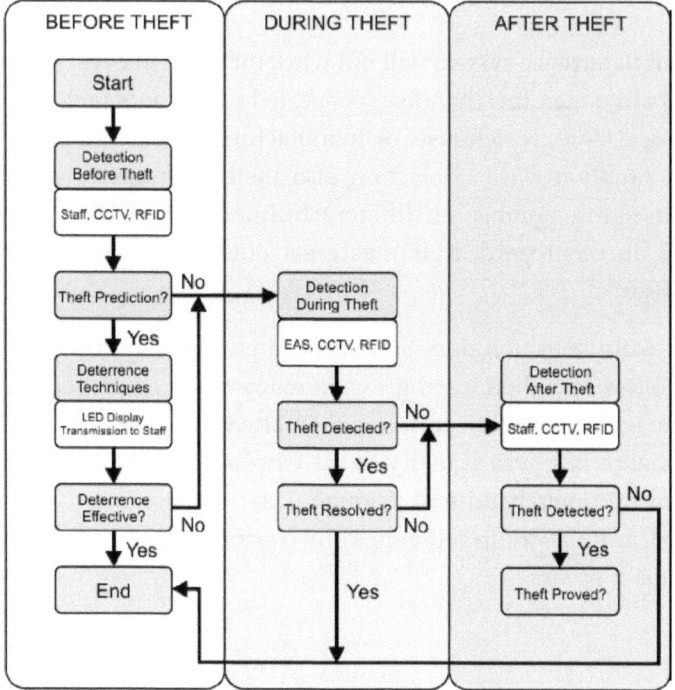

Figure 11-1 – The Three Conceptual Stages of Theft

RFID and Different Types of Theft

According to a previously published study, eight percent of all store customers shoplift merchandise while over 50 percent of employees at retailers steal from their employers.[27] In spite of these statistics, local police often consider shoplifting and internal theft soft crimes. Typically, theft from stores or warehouses is non-violent, in contrast to other crimes such armed robbery.

With scarce resources, law enforcement authorities place a higher priority on solving crimes that are more serious. For this reason, the control of shoplifting and internal theft becomes by default the direct responsibility of retailers and manufacturers.[28]

Few firms employ an integrated solution for theft reduction that includes real-time information on movement of goods through their entire supply chain. A recent study states, "at best company information on this problem (theft) is partial and incomplete and at worst, it is non-existent." [29]

To reduce theft it is necessary to have information about the scope and size of the problem. RFID has the potential to provide granular data to help determine 1) where along the supply chain theft occurs, 2) when it occurs and 3) how it occurs. This is an important first step in identifying the scope of the problem.

The remaining subsections study the applicability of RFID to prevent specific types of theft. By no means is this list comprehensive. Thieves are extremely flexible and change approaches very quickly. However, each known type of theft discussed below fits into the classification model appearing in Figure 11-1.

Prediction (Before Theft)

Theft Prediction is possible with RFID because it can work in conjunction with middleware to recognize patterns in real time and trigger deterrence measures. To be effective at theft prediction, the readers have to communicate with the EPC tags, feeding a constant stream of real time data to the middleware. This type of monitoring is effective in combating three important types of theft.

Open Pack Thefts

This type of theft occurs when thieves remove primary packaging from items before taking the item from the facility – e.g. retail floor, backroom, or warehouse. Theft prediction is possible at the shelf if the thief removes tagged items from the reader field, and the middleware senses the disappearances.

Package design plays a fundamental role in enabling RFID to sense this type of theft. Designs currently exist making it impossible to remove packaging without disabling the antenna, causing the signal to be lost. Another method to combat this form of theft is to mount the EPC tag on the item itself as opposed to the primary packaging. This ensures that the tag could be read all the way to the checkout area.

Sweep Theft

This type of theft occurs when a thief "sweeps" a shelf, taking large quantities of products at once. By placing readers on shelves, RFID can detect abnormal item activity. With real time detection, RFID can electronically alert the appropriate personnel within the store or facility whenever large numbers of items disappear simultaneously. On the retail floor, the middleware can be used to trigger an information display so that the would-be thief would know that their actions are being monitored.

Disabling Tags

Thieves employ several methods to disable anti-theft tags, including the application of a strong magnetic field close to the tags or by shielding items in a metal or aluminum foil-lined bag. Smart RFID retail shelves can combat this shoplifting method by constantly monitoring the items and detecting suspicious activity through the sudden absence of tag signals.

Another disincentive for thieves to destroy RFID tags is the removal of the item's legal identity. This will mean the item cannot be re-sold in the legal retail supply chain thus diminishing its value. Essentially, the item becomes "counterfeit."

In all of these cases, the thief has not attempted to leave a facility with the merchandise. Simply tampering with a product, or removing a large quantity from a shelf is not legally considered theft, although it is reasonably certain that this type of behavior is an initial step that will lead to theft.

Detection (During Theft)

In the event that predictive and deterrence measures are ineffective, RFID is able to enhance detection of a theft as it occurs. For detection, all components of RFID, including RFID, EPC, ONS, middleware, PML, and PML Servers, are required. The RFID system has great potential to detect the following types of theft.

Concealment

This occurs when a thief hides an item under a jacket or in a bag and exits a facility with the intention of stealing the item. Since RFID does not require direct line-of-sight, detection of unpaid items can still occur even when the object is not visible. During checkout, most retailers will either deactivate the tag or set the appropriate payment flag for the item. Therefore, if facility designers place sufficient distance between payment counters and exit doors, with appropriately positioned readers, store staff can identify stolen items and take appropriate action. Similarly, warehouse security staff can tell if unauthorized goods are leaving the facility.

As noted, confronting thieves is a dangerous practice. Though RFID could be used to detect a theft in progress, it is unclear if the technology will aid in apprehension of thieves.

Barcode Switch

Another common theft technique involves switching bar-coded pricing tags by taking a tag with a lower price and attaching it to a more expensive item. Retailers have attempted to fight this practice by putting two bar codes on a single item. Even if the cashier notices a switched bar code, the thief can claim it was the fault of the store. "If you do not conceal anything and are ready to pay for the article, it is hard to prove felonious intent." [30]

A similar technique involves thieves removing price tags and claiming the item was theirs. Unless security observes the individual removing the tag, it is difficult to counter the thief's assertion. Since the EPC is unique to the item, all relevant information can be obtained from the network to determine proof of payment, price, and status of the item. This will help eliminate the uncertainty associated with mixed tags. If a false EPC tag is used as a substitute, the serial number cannot be verified within the retailer's database signaling a potential theft in progress. The EPC can also be queried in the manufacturer's database to understand if the number is valid in situations where the product is located in a warehouse controlled by the manufacturer.

Collusion

Situations exist where internal employees collude with outside individuals to commit theft. For example, cashiers might register a product of lesser value, or not register the product at all by hiding the barcode with their hand while sweeping the product through the scanner. In the warehouse, a dockworker might collude with a trucker to load extra, uncounted goods onto a pallet prior to shipment.

RFID does not require line-of-sight reads, reducing the possibility of false reads at registers by covering the scanner as can currently be done with bar codes. Consequently, it will be much harder to violate the integrity of the checkout process. In the warehouse, items can be tracked in real-time, as they are loaded into vehicles. If employees carry radio frequency identity cards, an association of missing item to employee can be recorded and, over time, patterns can be established.

Intentional Undercount

During physical inventories, employees can underreport the actual number of goods, and then return to take the "lost items" after the accounting adjustments for shrinkage. When this theft technique is used, physical and perpetual inventories will be in agreement. The intentional undercount might appear when reconciling the receipt of goods versus sales and deliveries, but usually these discrepancies would be included in the overall shrinkage account – an example of why internal theft is sometimes difficult to measure.

RFID middleware is used by RFID to detect the movement of goods in real time while the EPC provides a method of accounting for transactions executed on an object-by-object basis. A result of this granular real time accountability is the continuous updating of material and information system data. There is less opportunity for physical and perpetual inventories to be mismatched through intentional undercounting. In the event that physical and perpetual inventories do not match, reconciliations can be made on an item-by-item basis and a trace can be placed on every unaccounted item.

Trash

One of the more unusual methods of stealing from a facility is to put items into trash for pick up later. RFID readers near the disposal area will help detect items that should not be in the trash and the middleware can then send the appropriate warnings to relevant personnel.

Proof (After Theft)

Depriving thieves of the ability to convert goods to cash is an excellent deterrent to theft. In the event that items are stolen, the ability to find out where they were manufactured and when they disappeared from the supply chain will not only provide retailers and manufacturers with valuable data as to weak points in their supply chain but also give law enforcement the opportunity to take action against the contraband.

Burglary

It is very difficult for any technological system to protect against forced entry or armed robbery. This is also true for RFID. However, the technology does provide an indirect means to counter burglary.

In many situations involving burglary, thieves convert stolen goods into cash through "black" or "gray" markets. Using RFID technology, an investigator could scan the EPCs of goods in these markets to determine their history. It might also be feasible to post a central list of EPCs for potentially stolen merchandise. This provides legal and corporate authorities with information to fight the unlawful sale of products through unauthorized markets. Slowing down the re-sale of stolen merchandise eliminates a major incentive for theft to occur.

Grab and Run

Brazen shoplifters will take items and rush out of the door. They rely on speed and surprise to escape security. The list of methods for this type of theft includes using emergency exits, running the anti-theft gates, and simply walking innocently out the store.[31]

The overall ability of RFID to prevent this shoplifting technique is minimal. As is the case with burglary, the unique EPC could potentially help law enforcement officials in determining the history of a suspect product if the items turned up in 'black' and 'gray' markets.

Fraudulent Refund

This technique involves a thief entering a store, taking something and proceeding straight to the returns counter. The idea is to create the impression that the thief came into the store with the item. While trying to return the particular item, the thief pretends to have lost the sales receipt or implies it is a gift. Store employees presently have almost no way to determine whether the item was part of a legal purchase.

A variation of this technique involves stores that have a "no refund without receipt" policy. In this case, the thief visits the returns desk with an unchecked item. When the clerk denies the request because of the lack of a sales receipt, the thief then walks out the door with the item completing the theft.

In both these circumstances, the ability for the returns clerk to trace the history of the particular item by using the unique EPC will help reduce the uncertainty concerning about whether the item has been stolen from the store.

Fraudulent Receipts

Shoplifters sometimes create a fake receipt and use it as evidence of a purchase if security or store staff approaches them. The thief enters the store with the receipt, goes to the shelf, and takes the corresponding item. Alternatively, the thief can purchase the desired item, leave the store, and then return to steal the same item repeatedly. If caught, the individual will use the receipt acquired earlier as proof of purchase.

The EPC's ability to uniquely identity the item means it can be matched against the financial transaction (and receipt) used to purchase the item. This level of granularity will help eliminate any uncertainty between the physical product and the accompanying receipt.

An Integrated Solution to Theft

RFID offers a comprehensive solution through a hardware/software infrastructure based on open standards. This is in contrast to current anti-theft applications that work independently using proprietary systems. As such, RFID is a flexible approach that can be adapted to meet the changing requirements of manufacturers and retailers. Summarized below are three additional technical attributes that make RFID an appropriate solution. Each attribute plays an important role in thwarting the different types of theft.

Embedding RFID Tags into the Product or Packaging

Tags embedded as part of the item or ingrained in the packaging will make it difficult for thieves to identify and remove the security tag. The embedded tag also prevents attempts to remove or switch price tags because thieves would have to remove the packaging to do so. Tampering with packaging is cumbersome, draws attention, and reduces the resale value of the item in secondary markets.

Expanded Scope of Coverage

With reader costs expected to drop to the $100 – $200 range, firms can place readers in shelves, aisles, shopping carts, and at exits and entrances. With the expanded coverage, internal thieves will have difficulty colluding with outside parties to steal from warehouses or by using backrooms and emergency exits to remove goods. Shoplifters will not be able to conceal merchandise and escape detection while leaving the store. While the long- term goal is ubiquity, the relatively high current price of readers restricts their use to a portal approach where important access points such as entrances and exits will be the optimal locations to accomplish a read event.

Reduced Human Involvement

Overall, the RFID system provides a dynamic platform for addressing different stages of theft. As more sophisticated applications that are developed at the Predictive and Deterrence stages, there will be an opportunity to reduce the need for human intervention in the resolution of theft. This will not only make for a system less prone to human judgment, but will help contribute to the important goal of a safer and less confrontational environment for employees.

Additional Benefits

Besides the immediate impact of reducing both internal and external theft, RFID technology provides several important long-term benefits.

Product Display and Store Layout

With RFID, managers can feel comfortable about open access to merchandise on the floor. About 40 percent of all purchase decisions rely on impulse buying, so allowing consumers to touch items will increase sales.[32] In addition, stores can optimize the display of goods for maximum merchandising impact.

Greater Control of Theft Prone Items

Hot products are items most desired by thieves.[33] "In many stores, 10% of the inventory yields as much as 40% of the total shrinkage experienced by a retailer." [34] Currently, security personnel develop hot products lists based upon perceived theft potential. RFID technology provides the necessary data to pinpoint the products vulnerable to theft and identify patterns if the hot products change.

Source Tagging

There are numerous venders of anti-theft tagging systems, each with proprietary standards. Manufacturers must tag the goods in accordance with anti-theft systems used by individual retailers. In addition, retailers often request several tagging configurations including dual tagging, tagging every other item, and no tag. These requests are an effort by the retailer to optimize the trade-off between the cost of the tags and the amount of theft prevented.

With RFID, there is a single open standard. While the primary purpose of RFID tags is to serve as an Automatic Identification and Data Capture (AIDC) technology for supply chain management, the tags also can be applied in a variety of anti-theft applications. The multi-functionality of RFID tags has the potential to reduce the requirement to have specific SKUs for individual retailers. This will have a positive impact on inventory levels.

Dynamic System

Internal and professional thieves rely on the fact that current security solutions are static. Though cameras, complicated packaging, and inaccessible items serve to deter thieves, all of these measures can eventually be defeated.[35] Once thieves learn the vulnerabilities, they are free to exploit these deficiencies at will. With no real time information about theft, it takes many months for manufacturers and retailers to realize if theft control methods have been compromised. Retailers must suspect the effectiveness of any anti-theft strategy that has remained the same for a long time.[36]

RFID, however, is a real time system that can constantly be adapted to meet changing needs. By receiving immediate feedback on theft patterns, security officials can find weaknesses and make changes.

Getting Ready for RFID as an Anti-theft Device

The task of making RFID work in practice is perhaps more daunting than the initial research on RFID tags or reader technology. As well, with any new approach there are many "buy-ins" that must happen at all management levels for manufacturers and retailers.

The vision of a total RFID solution for theft is conceptually sound. However, there will be a measured transition to practice as further technological development occurs. Engineers, information technology specialists, and security professionals must analyze numerous details of implementation along with financial justifications. Recent research work in the area of shrinkage takes an important first step toward the economic rationalization of RFID.[37]

Making implementation more complex, each market channel provides a different challenge for RFID anti-theft applications. However, manufacturing and warehouse environments present even greater challenges. The next two sections provide a brief analysis by market channel and manufacturing operation.

Retailers

Stores offer a clean environment for electronic equipment to operate. Smaller stores associated with specialty market channels often carry high value merchandise in a relatively small store space. In these situations, RFID applications for theft prevention can focus on a concentrated area for detection and deterrence. Several readers positioned at doorways, and the application of tags to merchandise, probably will reduce a large amount of theft and improve overall inventory management. With small store dimensions, more advanced applications become feasible at reasonable cost. Use of "absence of signal" to alert of a possible theft in progress has the best chance of working in a small space that is blanketed with readers.

For mass merchandise market channels, the large size of stores will be a prime determinant in the level of RFID application to restrict theft. Two strategies exist; 1) a store could reserve a small area for intensive RFID applications including automatic checkout or 2) management could rely on general tagging of merchandise with readers located at return areas, store exits or blind spots. The cost of readers will determine which strategy makes greatest sense.

Regardless of the size of the store, RFID theft applications will follow a similar architecture. Figure 11-2 shows one view of a complete conceptual diagram of RFID infrastructure for a store.

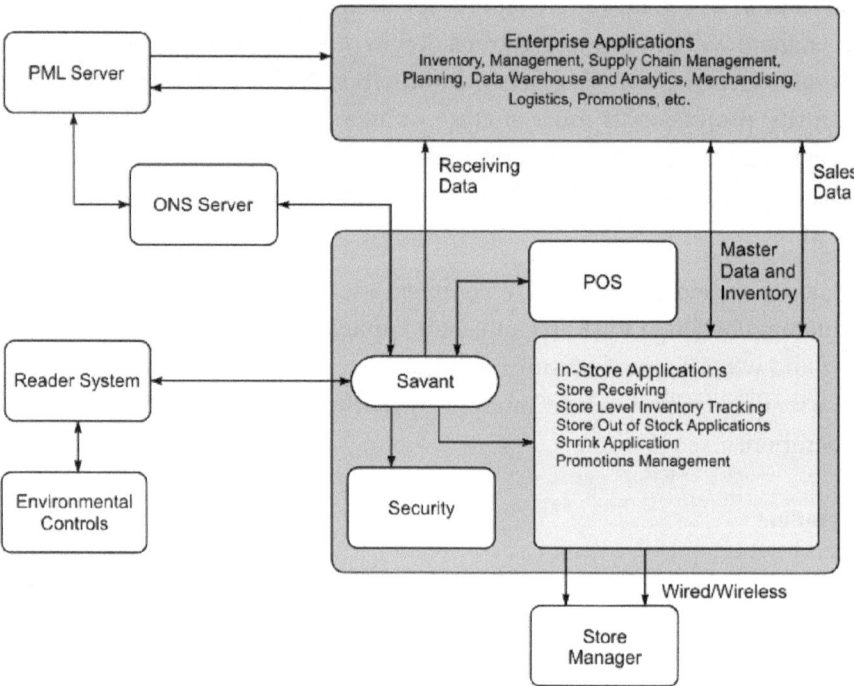

Figure 11-2 – Conceptual Applications Environment – Store

Adapted from IBM (Alexader, et al., http://www.autoidlabs.org/)

The key interface for store POS and corporate ERP systems is the RFID middleware. This means that RFID will require additional computing power at the store level. A network handles store-level data transmissions from the readers and hosts the RFID middleware. Remote servers handle PML files with a direct connection to the ERP system. The important issue with this architecture is that the middleware will plug into existing store POS systems. In this way, RFID is a decentralized system that can be adapted to individual store operations. The critical link is between the middleware and the ONS server. Likely, there will be a number of data transmissions between these two important parts of RFID. An interruption in transmission causes all theft monitoring to cease. Although this risk exists, it is a tradeoff with intensive investment in additional store level servers for PML and ONS. A decentralized architecture makes data maintenance extremely difficult to coordinate.

Manufacturing Plants and Warehouses

In contrast to stores, plants are a harsh environment. Even the cleanest operation is seldom free of dust particles that can cause difficulty around electronic equipment. Temperature and humidity vary a great deal and electrical utilities throughout manufacturing plants and warehouses often emit strong electromagnetic fields that can interfere with tag reads. Always present is the danger of damaging the readers. The manufacturing environment will probably always require greater maintenance and more attention to design of hardware that can operate in a number of difficult conditions. The widespread use of radio frequency communication with bar code systems gives some confidence about operating an RFID system in manufacturing environments.

The architecture for plant and distribution center applications (see Figure 11-3) of RFID is similar to the case for stores. From a systems standpoint, implementation of RFID in manufacturing and warehousing is simpler. Most firms already have extensive ERP systems in place. The previous use of bar code systems also serves as a base of expertise to build upon.

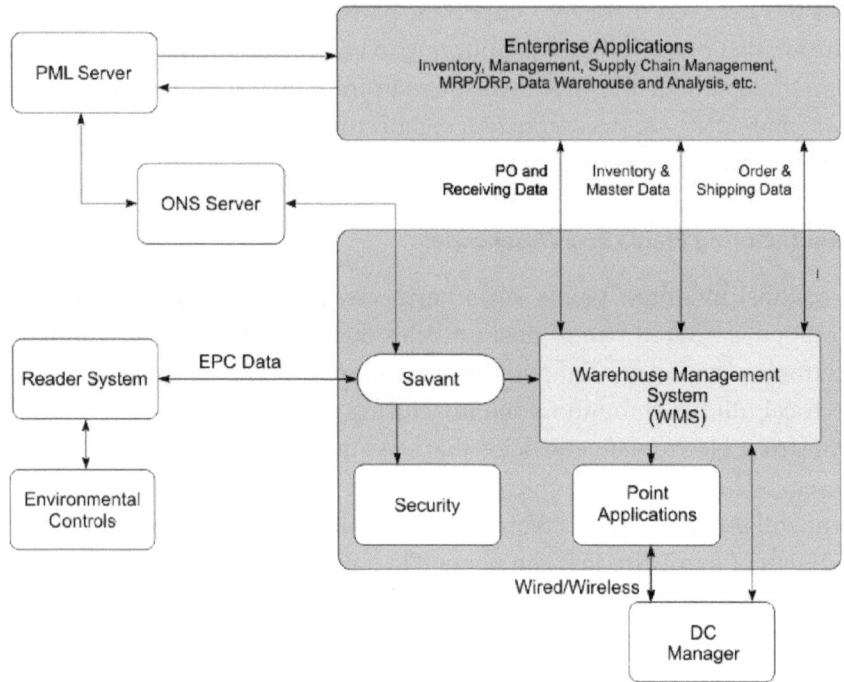

Figure 11-3 – Conceptual Applications Environment – Distribution Center

Adapted from IBM (Alexader, et al., http://www.autoidlabs.org/)

Following a similar combination of central and distributed data processing, RFID will have the warehouse management systems as a primary interface. With many warehouse operations having extensive computing capabilities, there should be less risk of system interruptions because all communications are self- contained. In the case of retailers, there might be one thousand stores to install RFID. The amount of installations would be much less for manufacturing and warehouse operations.

Implementation of RFID

Beyond issues of programming and IT infrastructure, RFID requires a detailed plan for installation. Currently, many stores and warehouses do not have the electrical utilities needed to support RFID. Readers in shelves and at exits must rely on power from ceiling electrical conduit pipes. Likewise, data transmission wires must link to networks. The option exists of achieving both power and data transmission links using wires located in floor channels. However, this requires extensive modification of stores.

Few other options exist to get electrical power to readers; however, data from readers can be transmitted by infrared signal to networks. Several companies currently provide this technology and it might be a better choice than extensive modifications to stores and warehouses.

False Alarms

The success of any new technology ultimately depends on its reliability. There must be a high percent reliability if a potential theft in progress causes alarm activation. False alarms are particularly disturbing to customers and employees alike.[38] One survey shows that 16 percent of people would no longer shop at a store if they were subject to a false alarm and wrongly accused of stealing.[39] Further, 50 percent of people surveyed indicated that high technology theft prevention devices make them feel uncomfortable.[40] Given this data, stores must be sensitive to concerns that customers might have about new technology.

Conclusion

RFID has potential to reduce internal and external theft within the supply chain by providing real-time information to security managers. Though RFID technology will never be a complete substitute for good procedures to reduce theft, it does provide an open standard for the theft prevention systems of the future. It is important to note that reducing theft requires a holistic approach that includes optimal business processes, qualified employees, training, technology, execution, store design, and the quality of information available.[41] With this view, a new technology such as RFID, used in isolation, will be less effective in practice as thieves learn ways to circumvent the system. This has been the case with past innovative technologies such as CCTV and EAS, which were initially thought to be a total solution to the problem of theft. [42]

Recognizing the potential for supply chain efficiencies, Wal-Mart and other retailers such as Tesco (U.K.) and Metro (Germany) have invested a great deal of time and money into developing RFID systems with the potential to reduce theft among other benefits. As of January 2005, Wal-Mart had RFID readers in 104 Wal-Mart stores and 36 Sam's Clubs. At that time, the company had received over 200,000 tagged cases and had 1.5 million EPC reads.[43] In total, 57 suppliers were using RFID technology, tagging inbound pallets and cases delivered to select Wal-Mart distribution centers.[44]

While this represents impressive progress, Wal-Mart, Metro, and Tesco have indicated that there remains much work to be done before RFID achieves its potential in a number of areas including theft reduction. Tag costs must decrease a great deal from current levels before tagging of a wide range of end items becomes feasible. At current prices, it makes sense to tag only high value items that are compact and prone to theft along with items where the case contains one unit, such as consumer electronics.

In addition, tag read rates must increase before a reliable theft control system based on RFID technology can work in practice. Though progress has been made concerning the read rate issue, there are still a number of variables that limit reliable reads in many different situations. To sum up the current state of affairs, Metro's Zygmunt Mierdorf has stated, "We are at the very beginning of the journey." [45]

As a concluding note, several researchers and consultants believe that process failures represent a much higher proportion of shrink than previously thought.[46] In this view, process failures cause shrinkage because of the inability to achieve precise measurements of inventory along with a lack of proper control of operations.

Support for this theory comes from the fact that almost all of the industry data on shrink is from surveys where respondents often rely on "guesswork and anecdotal evidence" to establish the causes for losses.[47] The suspect nature of the available data draws into question the true proportion of shrinkage attributed to internal and external theft.

Assuming that process failures will take on greater importance in the future as a target for management action, RFID continues to have the potential to play a significant role in the overall reduction of shrinkage. However, other management actions such as improved methods, training, and equipment also provide effective means to reduce shrinkage associated with process failures and theft.

The long-term viability of RFID technology as a tool to reduce theft and remedy process failures will ultimately depend on management choice, as the costs and advantages of other alternatives are considered in the context of the overall needs and goals of individual organizations.

CHAPTER 12

Defense: Improving Security and Efficiency

The problem of military logistics is as old as warfare itself. Supplying armies with provisions, arms, and military equipment requires a sophisticated level of coordination in terms of logistics and inventory control to ensure victory. An important aspect of military logistics includes obtaining accurate information on the location and amount of various supplies needed to conduct a campaign, which can often extend great distances into the hostile territories of opposing sovereigns. The course of history records many examples of reliable supply lines being the decisive factor for the success or failure of warfare.

In ancient times, some large foreign military campaigns took place even as measured by today's standards. Perhaps one of the greatest expeditions of antiquity involved a decisive battle that cast the fate of the Roman Empire. In 363 A.D. the Emperor Julian, who was accustomed to significant military success in continental Europe, fielded an army of 65,000 with the intention of attacking the forces led by the Persian King, Sapor.[1] Upon reaching the Euphrates river, the eastern most boundary of Roman domination, Julian assembled a fleet of eleven hundred ships containing "an inexhaustible supply of arms and engines [military equipment for sieges], of utensils and provisions" that included "vinegar and biscuit for the use of the soldiers."[2] The logistical support provided by the river-based armada of ships contributed to the early success of the campaign as the Roman soldiers won several important victories along the shores of the Euphrates.

However, as Julian moved farther into Persia beyond the Tigris and Euphrates rivers, he was forced to destroy his stores and scuttle much of the fleet of ships to prevent them from falling into the hands of the enemy and strengthening their position. This tactical move left his troops located in the hostile deserts of Persia with only 20 days of provisions.[3]

Julian's unanticipated death in battle resulted in an agonizing western retreat without the benefit of adequate provisions or the ability to live from the land, which had been burned by the subjects of the Persian King as a defensive measure.[4] The retreat, and the subsequent concessions made in a peace treaty with Persia by Julian's immediate successor, Jovian, weakened the empire and marked the beginning of decline. Historians attribute this battle, characterized by poor logistics and the unfortunate circumstance of Julian's death as the turning point in Rome's fortunes, although other factors such as political instability and religious conflict also contributed to the decline.[5]

History also records the early challenges of logistics for U.S. military forces as the country began to assert itself as a world power. During America's first major foreign military expedition, the Spanish American War of 1898, there were numerous reports of major difficulties in organizing logistics for the campaign. As an example in point, President McKinley initially decided at the last minute to increase the landing force destined for Cuba from 10,000 to 70,000 troops. However, just prior to the launch of the invasion, he discovered to his amazement that "there was not enough ammunition in the United States to keep such an army firing for one hour in battle."[6] After an urgent cancellation and a reduction of the landing force to 20,000 troops, the proper staging of supplies in the Southern U.S. still did not take place until one month later.

The logistical problems persisted beyond the initial delay. Upon embarking from Tampa to land on the shores of Cuba, an ambitious young cavalry officer, Lt. Col. Theodore Roosevelt, who was commander of a regiment named the Rough Riders, wrote in his diary "no words can paint the confusion."[7] As the fleet of ships prepared to depart for Cuba, "desperate quartermasters broke open dozens of unmarked railcars to see if they contained guns, uniforms, grain, or medicinal brandy."[8]

Amid the logistical confusion at Tampa, the commanding General ordered that no horses be loaded on the ships for lack of space. The Rough Riders, accustomed to cavalry drill, fought the entire war on foot. Thus one of the most famous battles in American history, the charge of San Juan Heights, became an infantry operation by default rather than a cavalry maneuver as romanticized in American folklore.[9]

In recent times, foreign military operations are more organized and this aspect of management no doubt has been the primary contributing factor to battlefield success. However, logistical difficulties continue to result in waste and disruption.

During the 1991 Gulf war, unmarked steamship containers became a concern as military planners had imprecise information about contents and arrival times. Between 40% and 75% of all containers shipped to Middle East depots had to be opened to find out what was in them.[10, 11] By one account "...some soldiers ate three breakfasts a day because they couldn't find lunch."[12]

In addition, a report from the General Accounting Office concluded, "$2.7 billion worth of spare parts shipped to the Gulf theater in 1991 went unused."[13] Without real time information about what supplies were in transit, planners often re-ordered supplies causing a large stockpile of unneeded materials to form at depots located in the Middle East. To cover the uncertainty of not knowing what was in transit to the Gulf, U.S. field commanders often ordered three times what they needed to improve their chances of getting supplies.[14] At the end of the war, the U.S. military destroyed many of these materials at great expense.

According to an initial report by the Department of the Army, the 2003 Iraq war also was characterized by logistical problems that affected operations.[15] During the military dash to Baghdad, forces outran their supply lines causing significant shortages in items ranging from tank engines to paper clips. In one case outlined in the report, the Third Infantry Division was within two weeks of halting operations because of a lack of spare parts. In another case, soldiers scavenged parts from abandoned Iraqi howitzers to keep their own weapons in working order.

These historical examples introduce the complexities of warfare and the importance of having real-time information about the quantity and location of supplies in support of armies. In contrast to applications in the consumer goods, agricultural, pharmaceutical or healthcare industries, the U.S. military supply chain offers an entirely different set of challenges for the application of RFID technology.

To ensure the readiness of the U.S. military to fight and win in any conflict, the Department of Defense (DOD) maintains a worldwide inventory of supplies and equipment valued at $67 billion.[16] Keeping track of this inventory is a large-scale systems problem with one of the most important inputs being real-time information about the location and availability of materials.

Achieving real-time information for a logistical network that contains a large variety of different materials and a great deal of physical movement requires the use of a unique identification system along with a massive information technology infrastructure. The development of RFID technologies presents an opportunity for the DOD to begin a change process that will alter the nature of planning and execution for military operations. This change will improve the effectiveness of American troops in a range of future combat scenarios that are likely to occur.

Information and Networking as a Future Strategy

Managing the widely dispersed inventory controlled by the DOD is difficult because the various military services and organizations that make up the U.S. armed forces use different automated supply systems. Adding to this difficulty, U.S. involvement in military conflicts is hard to predict and often occur quickly. For all of these reasons, the complexity of military logistics has reached an all time peak. Never before has it been so important for political and military leaders to know the exact level of readiness for the armed forces.

Recognizing the need for improvement, the DOD has initiated an effort to move the military out of the industrial age and into the information age. The goal is to acquire the ability to do networked warfare where information becomes as important as strength in force.[17] The military needs to be light, flexible, and agile to deal with future threats that could be drastically different from those faced in the past. It is becoming less likely that the strategic plan of the cold war era, which included preparations for mass infantry invasions and use of big-ticket strategic weapons such as Intercontinental Continental Ballistic Missiles (ICBMs), will be a politically viable alternative. Future world conflicts will increasingly be characterized by small clusters of combatants hiding out in "Third World slums, deserts, and jungles."[18] Massive buildups of supplies for foreign campaigns, which took 90–120 days in the case of the 2003 Iraqi war,[19] will no longer be quick or flexible enough to meet the challenge of asymmetric warfare where opposing forces are not equally matched and the rules of engagement are erratic. Creation of an effective organization to deal with this new threat of guerilla warfare is a formidable task involving 1.4 million existing active duty personnel, a civilian support structure numbering 700,000, and a large contingent of military contractors.[20]

As a first step in moving the DOD into the information age, planners are adopting a uniform identification standard for all supplies needed for military operations. By some estimates, this effort could save the DOD billions of dollars per year through elimination of redundant systems.[21] The capability to have real-time information about the location and availability of inventoried materials is critical for overall planning at the strategic, tactical, and operational levels. In addition, it is critical to maintaining troop readiness.

Real-time visibility throughout the military supply chain improves the chances that soldiers will have the proper equipment, at the right time, to ensure success when entering battle. This reduces the likelihood of casualties and improves the morale of the military force. Soldiers going into battle deserve the assurance that important supplies will be available when needed.

To guide the military in issues that involve long-range strategic planning for the new realities of American security, the Joint Chiefs of Staff issues a report once every five years that looks into the future and establishes goals as part of a continuous improvement process for military effectiveness. The most recent report, titled Joint Vision 2020, is a significant document that recognizes the role of logistics and supply chain management in the success of future warfare. An important operational concept of the report is Focused Logistics, defined as:

> "The ability to provide the joint force the right personnel, equipment, and supplies in the right place, at the right time, and in the right quantity, across the full range of military operations. This will be made possible through a real-time, web-based information system providing total asset visibility as part of a common relevant operational picture, effectively linking the operator and logistician across Services and support agencies."[22]

Building a real-time, web based information system that provides total asset visibility depends on the ability to link physical items to the Internet using low cost RFID tags that can uniquely identify individual objects. For an organization as large as the DOD, it is critical that a universal standard be adopted so that interoperability exists within all branches of the military. This ensures maximum supply and equipment availability to the soldiers and offers the opportunity for joint procurement of common items. Interoperability and joint procurement reduces the cost of military operations.

As important as a universal standard for identifying materials and equipment has become for the military, the scope of the DOD does not include development of comprehensive standards for identification. Rather, the DOD depends upon civilian groups such as non-profit organizations, academia, and industry to develop the standards. Once fully tested, the military then adopts these standards, making the civilian-military interface reflect the common practices of business.

DOD Supply Chain

In many ways, the DOD supply chain is similar to those of commercial suppliers because many of the products and supplies contained within the DOD supply chain are also available commercially. However, differences in optimization criteria lead to a number of characteristics that set the DOD supply chain apart from the commercial supply chain. For the most part, civilian logistics managers use cost and customer service as the main optimization criteria. In contrast, officers and civilian personnel planning military supply chains have much different objectives. Some of the most important of these differentiating characteristics follow.[23]

Readiness. The primary purpose for optimizing the military supply chain is to enhance readiness for war. Knowing the location and status of all materials needed to support operations is an essential component of readiness. In some regards, this is similar to customer service as measured in business. However, the outcome of failure in meeting the standards needed for military readiness is battlefront failure and the loss of lives.

Long supply lines. War is an international activity, which means that lines of supply to support operations are long. Without automatic identification technology that provides real-time visibility of items moving from the suppliers to the front-line troops, it is extremely difficult to maintain accurate knowledge of supply-chain-wide inventories. While it is true that international companies also face long supply chains, few rival the complexity of the U.S. military. Another differentiating factor is that military supply lines within hostile territories must be secured against attack by the enemy.

Variety of items. Military operations require a large number of items ranging from everyday supplies to food and clothing to specialized equipment. Different categories of items have different standards for inventory accuracy and visibility.

Unstable demand. Military demand is often variable and unpredictable because conflicts can happen anywhere in the world at any time. When a conflict occurs, demand for supplies increases dramatically and existing stockpiles of materiel are depleted quickly. Accurate inventories are critical to maintaining readiness in the presence of variable demand.

Moving end-points. The end, or destination, points of the military supply chain generally move forward with advancing troops and are either terminated or transformed, creating additional difficulties for transportation and inventory management.

Priority. The military supply chain operates on priorities set by unit commanders based on urgency of need.

Equipment reliability and maintenance. Military operations take place in all types of environments and on all kinds of terrain. Under battle conditions, it is important that all identification technologies work effectively and that system maintenance is minimal.

Detection. In a theater of operations, the military must always be careful not to divulge information about its position that would be advantageous to the enemy.

As an example to highlight the overall complexity of the military supply chain, Table 12-1 shows the wide range of materials needed to support the U.S. Army.[24]

Table 12-1 – Procurement Classes of Materials and Equipment

CLASS	DESCRIPTION
I	Subsistence and commercially bottled water.
II	Clothing, individual equipment, tools, tool kits, tents, administrative and housekeeping type supplies, as well as unclassified maps.
III	Petroleum, oils, and lubricants: includes bulk fuels and packaged products such as antifreeze.
IV	Construction items, including fortification and barrier material.
V	Ammunition.
VI	Personal demand items (nonmilitary sales items) and gratuitous health and comfort pack items.
VII	Major end items, such as launchers, tanks, mobile maintenance shops, and vehicles.
VIII	Medical supplies, including repair parts for medical equipment.
IX	Repair parts and components required for maintenance support of all equipment.
X	Material to support nonmilitary programs, such as agricultural and economic development.

Each class of material listed in the table receives a rating in terms of importance for a specific type of operation. When doing planning, the military builds various combinations of the classes to support operations in peacetime and during war. The characteristics of each class, such as cube, weight, shelf life, and level of hazard to personnel become important criteria in logistics planning. In many respects, the military classification system is similar to the freight rating system used in commercial practice.[25]

The problems that have resulted in the past from the special characteristics of the DOD supply chain often were exacerbated by poor inventory visibility. The use of RFID systems that are customized to accommodate the unique aspects of the DOD supply chain can significantly reduce the recurrence of these problems. Some general areas where inventory visibility is important for military operations include:

Inventory Management

Recent analyses have found that faulty inventory records often result in miscalculated order quantities.[26] In addition, shipping delays create uncertain transit times. Faced with poor supply chain visibility, military planners have no choice but to over-order in an attempt to compensate for uncertainty. This leads to invalid priorities, excess inventory, and bottlenecks in transportation. Accurate, real-time inventory management throughout the supply chain would improve visibility and reduce over-ordering.

Repair and Maintenance

Any large-scale repair operation is complex because it is difficult to predict demand for spare parts. In military repair operations, expensive parts are given high priority and customer wait-time is usually very short. However, inexpensive parts are often critical to completing a repair. These parts are usually assigned a lower priority, which often causes them to be delayed in shipment. In turn, this causes delays in the entire repair cycle.[27] Military planners often increase the total fleet size to compensate for lengthy repair times.[28]

In addition, the long life cycle for many weapons systems used by the military means that there are often several different versions of the same part depending on how many engineering changes have been made through time. The existence of several different versions of the same part number causes confusion in maintenance operations as technicians attempt to determine if one version of a part fails more frequently then another version. Simply locating all the versions of a part throughout the entire military supply chain can be a major operational bottleneck.

Readiness and Mobility

Combat forces must be ready to engage in a conflict, and they must be able to move to the conflict location quickly. Troop readiness is determined in part by equipment readiness, and equipment readiness hinges on proper repair and maintenance.

Mobility is determined primarily by the quantity of materiel that must be moved and the number of transport vehicles available to carry it. In general, the smaller the inventory required to travel with a force, the greater its mobility. Accurate data on inventory quantities and locations enables logistics support systems to transport a greater quantity of items, because there is less need to transport extra stock to cover for the uncertainty that results from inaccurate inventories.

Tracking

The lack of a single, standardized RFID system severely limits the tracking of assets as they move through the supply chain from the supplier to the troops. Similarly, the visibility of objects flowing back through the supply chain is limited. The inability to track individual items negatively affects all supply-chain-related applications, including repair and maintenance, identification of failure-prone parts, and the ability to perform predictive maintenance.

System Improvement

In response to recognized problems with asset visibility throughout the military services, DOD has made significant investments in research and development of RFID systems that will improve security, cargo visibility, hazardous materials (HAZMAT) recognition, inventory management, product tracking, and quality control.

Security.[29] Automatic Vehicle Identification (AVI) is a project to enhance security at access-control points. The Army has hired Transcore, Inc., to test access control at Fort Monmouth, New Jersey, using passive, ultra-high-frequency (UHF), "eGo" wireless RFID tags.

Testing of eGo tags began in November 2002. Vehicles with proper security clearances are equipped with eGo tags on their windshields. As vehicles approach the entrance to the fort, they encounter a simple tilt-arm gate. An RFID reader scans the eGo tag, and the gate opens. The car then proceeds to a common access reader, where the driver is identified using established, non-RFID procedures.

Technology used for this test includes the thin eGo windshield tag and the eGo 2210 reader. The tag has 1,024 bits of memory, is tamper resistant, and can withstand extreme temperatures, sunlight, humidity, and vibration. Since this is passive tag technology, the read range is about 5 meters. Approximate cost of the tag is $10.

Cargo visibility.[30] DOD and Savi Technology are collaborating on two projects, Smart and Secure Tradelines (SST) and Total Asset Visibility (TAV). With SST, tags are placed on cargo containers before they are shipped. The tags record any activity during transit, such as nonconformance to security measures, and make this information available on arrival of the containers at a port. The tags also include detailed information on what is inside the containers.

TAV was created by Savi Technology to track cargo containers and record their location at any time during transit. The system is based on Savi's Universal Data Appliance Protocol, which allows integration of devices such as RFID and global positioning systems.

HAZMAT recognition.[31] The Defense Logistics Agency has organized a test of Advanced HAZMAT Rapid Identification, Sorting, and Tracking (AHRIST). Currently, no system exists to alert receiving personnel automatically before HAZMAT arrives. The objective of the AHRIST project is to track HAZMAT through the supply chain, and to give notification similar to an advanced shipping notice (ASN) transmitted using electronic data interchange or some other computer-to-computer communication protocol. The Micron Technologies tags used in the test have a read range of 10 feet and 128 bits of storage.

Product tracking.[32] DOD is working with several companies on materiel-tracking applications. In 1999, Symbol Technologies was awarded a 5-year, $248 million contract for RFID technologies and services. Projects under this contract include tracking materiel deployed throughout the world and advanced identification of military personnel through ID cards containing RFID chips. The tracking system uses NATO (North Atlantic Treaty Organization) stock numbers and can distinctly identify 1.8 million line items. The computer system interface uses the IBM ES9000 Series mainframe to run Mincom's Management information System. The system tracks goods received, performs spot-checks, and notes other factors such as batch number, shelf life, expiration date, and reparable or non-reparable designation.

Quality control.[33] Quality control of meals, ready to eat (MREs), is currently being tested at the Army Soldier Systems Center at Natick, Massachusetts. Hardware and tags developed by Savi Technologies are used to inventory containers of MREs at supply points. Low-cost passive and semi-passive RFID tags developed by Alien Technology are being used to identify MRE cases and pallets. This project relies on the MIT Auto-ID Labs' technology to track shelf life and the environmental conditions (temperature, humidity, and vibration) under which MREs are stored.

Ordnance.[34] An important part of military operations, ordnance represents a large inventory that must be accounted for on a periodic basis. The U.S. Army has 5 million ammunition pallets worldwide, with about 10,000 earth-covered storage bunkers located in the U.S. Some of these bunkers are enormous, covering 100 square miles underground. Complicating matters, the product line is a broad category with a number of different stock keeping units (SKU's). This inventory also includes ammunition that is expired or no longer in working order. It is important to ensure expired ordnance never enters the military supply chain supporting active military operations.

To eliminate manual counting, which includes accounting for stock (product), lot, and serial numbers, the U.S. Army has implemented a new system to replace the existing use of bar codes. However, instead of implementing RFID technology, the U.S. army decided to use a system that utilizes two-dimensional (2-D) bar codes. The new system reduced the amount of time required to count the inventory and "reconciliation costs have dropped from about $1.5 million per depot to tens of thousands of dollars."

The choice of 2-D bar codes as opposed to RFID technology is important in that the electromagnetic waves emitted from tags and readers might detonate some types of ordnance. Beyond the issue of safety, the U.S. Army also cited that lack of infrastructure within bunkers to accommodate RFID technology as an important factor limiting its use in practice. Most of the bunkers are little more than holes in the ground with few utilities.

The 2-D bar codes can hold 100 times the information as compared to a conventional bar code, allowing the stock, lot, and serial numbers to be stored in readable format on the pallet. Though this method requires line of sight scanning similar to bar code readers, and simultaneous reads of many different pallets are not possible, 2-D does take advantage of the ability to do unique identification of individual pallets. In this way, the example of automating counting processes for ordnance demonstrates the flexibility of different approaches beyond RFID technology for achieving unique identification under extreme conditions.

How RFID and DOD Come Together

The most important element missing from current DOD testing of RFID systems is standardization. A standard system for RFID across the entire DOD will facilitate inventory management and related applications, thereby creating increased readiness at a reduced cost. Several future possibilities exist to utilize current open standards that are being administered by EPCGlobal.

Tracking. Using mass serialization and the Office of NATO Standardization database, the EPC system will allow for real-time tracking of supplies with a single technology. A standardized inventory management system will give visibility of the location of spare and repair parts. Maintenance and repair then will be more efficient, and applications such as predictive maintenance will be possible.

Product identification. The EPC is a unique identifier that points to a database holding all information about an item. The current Universal Product Code holds only limited information about a product and its manufacturer. With the EPC system, military planners anywhere along the supply chain will be able to access detailed information, including suppliers of each component of the item, transportation methods, and environmental storage conditions for the item throughout its lifetime. With more information about an item, military planners will be able to make better decisions.

Military-civilian interface. By using the same standard as industry, DOD will be able to communicate with commercial vendors and have direct visibility of inventories at civilian locations. An active military-civilian interface also will give vendors and military planners the opportunity to collaborate on ways to enhance readiness for war.

As an example, precise inventory levels by version will be possible with the EPC system. Under the current system that uses bar codes, a part number is not unique because a new number is not assigned each time an engineering change is made. Therefore, an inventory of spare parts for equipment that has a long life cycle often includes many different versions of a part as engineering changes occur over the life of the equipment.

Civilian warehouses will be able to assist the military in stockpiling enough supplies to sustain several simultaneous war scenarios. This could take the form of maintaining "warm" inventories that are reserved for military operations, yet continue to cycle into normal shipments. This practice would reduce losses resulting from exceeding shelf-life limits. Using inventory pooling between civilian and military organizations would significantly reduce waste and improve readiness.

Predictive maintenance. The military currently employs a preventive maintenance policy for complex equipment. This means that regularly scheduled overhauls take expensive weapon systems out of operation for long periods. This policy is not efficient. With the serialization capability of the EPC, the history of every service part can be stored in a database. The history is a vital piece of information in predicting failure. Rather than scheduling overhauls on a periodic basis, maintenance could take place when a military planner sees that a part is likely to fail. Service parts could be pre-ordered. Instead of ordering a larger, more expensive system part, such as an aircraft engine, component parts that are likely to fail, such as water pumps, could be identified and stocked. Vehicles would operate at their maximum efficiency, which would result in reduced total life-cycle costs.

Budgeting. The real-time inventory information provided by the EPC system would improve the accuracy of budgetary decision-making. Current budgetary decisions are sometimes based on faulty information about inventory levels and troop readiness. The EPC system would reduce uncertainty caused by counting or timing errors in the information used for budgetary purposes.

Table 12-2 provides a summary of specific RFID possibilities.[35] Boxes marked "RFID" suggest that identification tags would improve efficiency in that area. Boxes marked "Auto-ID" indicate areas in which the networked EPC system would be substantially more beneficial than a proprietary RFID system. The boxes marked "Being tested" indicate that RFID technology is currently being used or tested.

Table 12–2 – Implementation Opportunities for RFID in the DOD Supply Chain

	Weapons/ Machines	Ammunition	HAZMAT	MREs	Everyday Supplies	Shipping Port Containers	Personnel
Track	Auto ID	Auto ID	Being tested	Auto ID	Auto ID	Being tested	Being tested
Shelf Life-Product Information	Auto ID	RFID	Being tested	Being tested	RFID	RFID	
Inventory Management	Auto ID	Auto ID	Auto ID	Auto ID	Auto ID	Auto ID	
Recall	RFID	RFID	RFID	RFID	RFID	RFID	
Security	RFID	RFID	RFID	RFID	RFID	RFID	RFID
Military-Civilian Interface				Auto ID	Auto ID		Auto ID

Increasingly, success in warfare depends on accurate information on the identity and location of parts and systems. This is true not only for battle-field operations but also for the support functions that must get supplies to the right place at the right time. In the future, DOD will be expected to enhance readiness for war while minimizing procurement costs. The RFID technology will play an important role in this enhancement by providing open standards for both DOD and industry while creating unprecedented total supply chain visibility.

Conclusion

Since 1994, the U.S. military has issued contracts equaling $280 million dollars to implement proprietary RFID technology for tracking cargo containers as part of the DOD's TAV network, along with tracking other military supplies.[36] This is a large task in that the military must manage over 270,000 cargo containers needed to ship supplies around the world.[37] An early innovator in RFID technology, one of the first recorded applications of proprietary RFID technology in the military involved the tracking of air shipments to Bosnia in 1996.[38]

Because early pilot projects demonstrated the practical advantages of RFID technology for military supply chains, the DOD issued a mandate in October 2003 that requires all of its suppliers to use passive RFID tags on cases and pallets by January 2005.[39] NATO has also announced plans to implement a pilot RFID system to manage the flow of military supplies.[40]

The application of RFID technology by the military has not been without problems. In one case where a proprietary RFID system was implemented, the military ran out of serial numbers for tags.[41] This "required the Defense Department to update 20 major logistics management information systems and more than 1,500 RFID workstations worldwide in the past year." [42]

There have also been anecdotal accounts of high maintenance costs required to operate a RFID system. By one account, annual maintenance costs equal 30% of the installation cost.[43] Readers located in military warehouses are particularly susceptible to damage by forklifts.

Further, the DOD has its own numbering system for procured items called the Unique ID (UID). Since the announcement in 2003 requiring all vendors to tag items sold to the military, there has been confusion concerning if the military will require the UID or the EPC to be included on tags. In April 2004, there was clarification that the military will, "in many cases" accept the EPC as a replacement for the UID.[44] There have also been discussions of embedding the UID as part of the EPC.

As a final note, there are reports the Pentagon is planning to invest billions of dollars in a program called Future Combat Systems, with the broad goal of creating a 21st century military force of automated equipment for warfare that will be controlled by a "network of satellites, sensors and supercomputers." [45] In such a system, RFID technologies will play an important role in establishing the infrastructure for real-time information about the location and quantity of supplies needed for combat.

Even as the art and science of warfare advances to an era of automated combat where machines rather than humans engage in battles, logistics remains as an important element needed for success. In this regard, little has changed since the first organized armies of antiquity marched against opposing sovereign nations. The realities of warfare in terms of logistics are eternal to the history of humankind.

PART III: CREATING BUSINESS VALUE

CHAPTER 13

The Role of Data
in Enterprise Resource Planning

For nearly all supply chains, Enterprise Resource Planning (ERP) plays an important role in coordinating the various activities and business processes inside individual firms.[1] An enormously complex system, ERP depends on data to accomplish planning and scheduling tasks of great importance to the management of modern manufacturing. Increasingly, many service-oriented firms are also adopting ERP.

A flexible approach to coordination, ERP has a history that traces to the emergence of commercial computing systems during the early 1960s. While manufacturing depends on more than planning, scheduling, and control, ERP is an important element of manufacturing strategy, which has historically been comprised of three parts:[2]

1. Vertical integration to reduce cost and ensure raw material availability
2. Creating new products that are superior
3. Achieving dominant market share to take advantage of the economies of scale

While this strategy has worked well for many years, the increasing importance of the customer has driven manufacturers in new directions. Globalization, outsourcing, proliferation of stock keeping units (SKU's), and shorter life cycles[3] have brought customers expanded product variety at low cost, along with new types of supply chain complexity. For example, a typical third party warehouse in Ohio carries goods "made in 120 locations around the world."[4]

Adding to this complexity, the scale of data capture needed for input to ERP systems is likely to expand dramatically once the EPCglobal Network and RFID technology are operational. A large consumer goods firm might handle 60 billion tagged end-items per year, with each item being read an average of five times as it moves through a manufacturing facility or ware-house.[5] Given this additional volume of data, ERP systems will need to evolve in new ways to handle the EPC.

This chapter examines the role of the EPCglobal Network and RFID tech-nology in capturing input data for ERP systems. The implications of RFID technology affect several important areas including data interfaces, bill of material structure, accounting, and the treatment of capacity in material requirements planning. All of these are new developments that will change the nature of ERP in the years to come.

The Value of ERP [8]

In many respects MRP, the subsequent development of manufacturing resource planning (MRPII), and ERP, represent increasingly sophisticated databases that over time have improved tactical and strategic business planning. Essentially, ERP serves an "uncertainty absorption" function.[6] It is impossible to know with certainty all future outcomes that might occur for a business. However, with enough data and proper methods of analysis, reasonable projections of future outcomes become feasible. Having data allows for the possibility of calculating risk, where several different out-comes are possible, and a probability calculated from the data can be assigned to each outcome.[7]

The crowning achievement of ERP systems in practice is that business deci-sion making has moved from an *uncertainty basis* where no comprehension of risk exists, to a *risk basis* where ERP serves the important function of mit-igating uncertainty. The result: Much more effective business decision-mak-ing based on rational analysis of data available rather than pure conjecture. With the established success of ERP in practice, it is realistic to begin think-ing about what changes in information technology will further enhance ERP, thus reducing even more uncertainty within business planning. Since ERP is at its essence a data management tool, it is reasonable that any advancement in the way that data is obtained, organized, and employed will have a significant impact on the structure of ERP software.

Data and ERP Systems

Accuracy of data has been an important goal since the inception of ERP. Early efforts focused on improving the accuracy of the bill of material

(BOM). In the past, popular management programs such as Class A MRP II and cycle counting became important in helping practitioners get the most benefit from these systems. With the perfection of the BOM approach, emphasis has shifted to raw data accuracy as a means of further improving the overall results of planning.

Many firms have benefited from bar codes as a means of automatic data capture for raw materials, work in process (WIP) and finished goods. The use of bar codes has drastically reduced the amount of labor needed to conduct many basic business transactions. At the same time, bar codes have also improved data accuracy by reducing human input for data entry. Table 13-1 summarizes the progression of data entry methods associated with ERP and its forerunners.

Table 13-1 – Data and ERP

	MRP(1960s)	MRPII (1980s)	ERP (1990s)	ERP + RFID (2004)
Data Capture	Manual	Barcode + Manual	Barcode + Manual	RFID
Data Type	SKU code	SKU code	SKU code or item serial number	Mass serialization a serial number for each item or component
Pro/Con	Improved planning capabilities – limited data available, accuracy problems	Speed collection of data and improved accuracy, Batch mode – delays in updates	Standardized collection of data, some lot control – limited serial number control, lack of middleware, mature technology	Granular data at serial number level, middle ware to manage serial numbers, common standards, real time – initial stages of development, technology to read tags must be refined

In the case of bar codes, data is gathered through close proximity optical scanning. Updates to ERP occur in batch mode for the most part. While this approach increases the amount and accuracy of data available for ERP calculations, there are several limitations to bar code data capture systems. The biggest drawback affecting ERP is timeliness of inputs because of the difficulty in configuring true high-speed, fully automatic data collection points.

In spite of the success of the bar code, the goal of tracking items through an entire supply chain with 100% inventory accuracy remains elusive. This type of effort represents a huge challenge to current information technology infrastructures that are a critical part of ERP. In the future, automated methods of planning and control within manufacturing and service operations, and entire supply chains, will depend on accurate, real time information and unique identification of individual objects. Because manufacturing systems are in constant flux, data accuracy is not just a function of having the correct value, but of having the correct value at the correct time to reflect the proper state of the system. Accurate data that is old is of no use in a dynamic system.

RFID technology offers the potential to increase by an order of magnitude the amount, accuracy, and timeliness of data within businesses by blanketing the supply chain with readers. With RFID, real-time streaming data, filtering, processing, and response are possible.

With this in mind, it is no surprise that an independent online survey conducted by APICS (the American Production and Inventory Control Society) supports the importance of data accuracy. When polled about relevance and main goal RFID technology, 55% of respondents indicated improved inventory accuracy was the most important objective. The total results of the poll are as follows:

RFID Poll (spring 2004)

What is your main goal in implementing an Auto-ID solution?

Improve inventory accuracy	55%
Trading partner requirement	13%
Increase inventory turns	10%
Reduce out-of-stock situation	9%
Enhance supplier relationship	9%
Improve fill rates	4%
Sample size	*658 respondents*

Thinking beyond the utilization of real-time data, RFID offers other opportunities to capture detailed data about objects within a supply chain on a scale never before experienced in commerce. However, organizing EPCs represents a challenge requiring significant changes to ERP systems.

Organizing Data from the EPC

Properly capturing EPCs without the capability to use the data in a timely manner, because of the lack of organizing software, is a serious issue given the volume of EPC data anticipated. Having the right data means little if it cannot be properly applied within ERP at the right time.

Though it is early in the development of RFID technology, it appears ERP will play an important role in managing the EPC data needed for supply chain wide visibility. The EPC, a fundamental aspect of RFID Technology, provides the capability for unique identification of trillions of objects. Unique identification on this scale results in useful information for track and trace, and the authentication of objects located anywhere in a supply chain. However, managing serial numbers for trillions of objects presents a difficult challenge for current ERP systems to handle. As a result, there will be a measured transition from lot control, currently available in some ERP systems, to serial number control enabled by new software concepts such as the Transactional Bill of Material (T-BOM).[9]

With the T-BOM approach, serial numbers contained in the EPC are organized to provide the history of movement for an item (pedigree information), a schematic of the serial numbers for all components contained in the finished item, and a mechanism to allow a query for authentication by any party within a particular supply chain. This is accomplished through sophisticated database technology that utilizes EPC information gathered from the middleware interface to RFID.

The T-BOM represents a new generation of software intended to enhance system integration as RFID technology begins to take hold in industry. Since current ERP systems use only lot control for tracking and tracing, it is important to add capabilities that handle EPC data so that it can be queried and communicated as needed. Without these types of new structures to enhance ERP, there will be much less effectiveness in using data from RFID technology.

Besides tracking, tracing, and authentication, serial data on components opens new possibilities to gain insight into complex operations. There are many situations where lack of detailed information leads to ineffective supply chain management. For example, difficulties with management of versions is a common problem in the capital asset industries where service parts for long life cycle items such as aircraft frequently undergo modification and redesign midway through the life of the asset.[10] With most part numbering systems, different versions of a service part cannot be identified, inventoried, traced or tracked. In situations where there are large networks that do maintenance of deployed assets, such as airbases in support of combat aircraft, knowing the exact version of a service part in inventory is essential to providing high levels of service and readiness. In addition, the ability to track failure rates by serial number (version) is also critical to understanding overall reliability as service parts move from manufacture, to distribution and finally to installation and use.[11]

There is no question that RFID has great potential to provide detailed data about objects within a supply chain. The data capabilities of the technology also allow other possibilities such as a change in the algorithmic structure of ERP. The next section explores just a few of these possibilities.

Capacitated Planning and Automated Scheduling

One of the most basic processes of ERP is planning and scheduling. Figure 13-1 provides a conceptual overview of the various planning and scheduling functions common to all ERP systems.

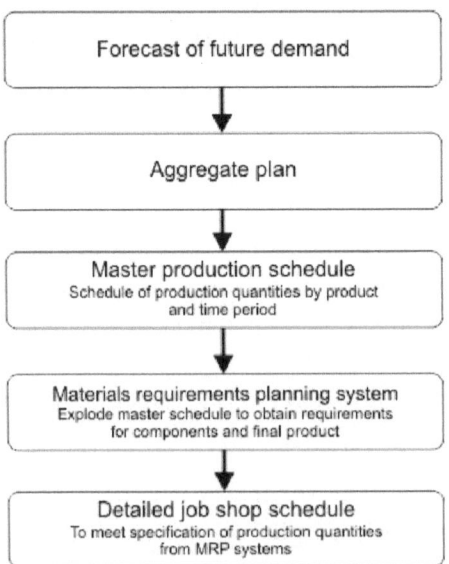

Figure 13-1 – Hierarchy of Production Decisions

Adapted from Nahmias (1993) [12]

Two aspects of RFID technology have the potential to change the way that practitioners use ERP for planning and scheduling.

First, the ability to have manufacturing plant and supply chain wide visibility of objects identified with the EPC allows for large amounts of information and executable instructions to be assigned to an object.[13] An example that has been in application for several years involves attaching an electronic tag to a component that is work in process (WIP).[14] As the component moves through different manufacturing stages, the tagged item is scanned and instructions are downloaded from databases into computer numeric control (CNC) milling machines that automatically cut the component to exact specifications. When the component moves to the next stage of manufacturing, another scan takes place and a new set of instructions are loaded into processing machines. It is even feasible that a queue of tagged parts for an individual work center could be scanned simultaneously to identify important information for adjusting work center priorities. In this manner, detailed day-to-day shop scheduling and management of instructions become automated processes.

With this level of control, there are almost unlimited opportunities to improve information handling and automation within manufacturing plants. The opportunity also exists to increase the level of automation across entire supply chains so that a component manufactured at one plant can be transferred to another with the knowledge that all relevant information and manufacturing instructions are attached to the component and can be processed automatically. The open standards and protocols are an important feature of the EPCglobal Network that allow for this type of information transfer and communication within the supply chain.

The second important aspect of Auto-ID technology that will change the way planning and scheduling is performed within ERP involves the continuous flow of data. A well designed RFID system is always "on." With this improved sensing capability, critical subsystems of ERP will have accessibility to more data for scheduling calculations. Given real-time data, new possibilities exist to apply advanced algorithms such as math programming and heuristics in every practical aspect of planning and scheduling.

One of the most important goals of manufacturing is the management of capacity utilization. Several ERP subsystems are crucial in achieving this short and medium term goal. The master production schedule, the MRP system, and the detailed shop schedule all visualized in Figure 13-1 are the current tools within ERP to manage capacity. For many years, all of these systems assumed infinite capacity when doing planning and scheduling.

This assumption, though widely recognized as an important weakness, reflected the reality that in many cases data did not exist to support advanced finite planning and scheduling. Planners have spent untold hours manually balancing production to meet available capacity. When the problem could not be solved manually, due dates were not met and customer service suffered.

Beginning in the mid 1980s, the advent of microcomputers resulted in the introduction of master scheduling software that accomplished capacitated planning and scheduling for end items. These software packages existed outside of ERP systems and required significant integration to achieve operability. During this time computer spreadsheets began to be used as a powerful means to build models and do finite capacity scheduling for end items. [15, 16, 17, 18]

However, achieving capacitated planning and scheduling for a single level, finished good, is far easier than achieving the same task for dependent demand (MRP). In this case, the consideration of capacity constraints and cost optimization must take place through multiple levels for the BOM. Manufacturing multiple complex end items at a single facility adds to this complexity.

MRP has been singled out by managers and academics alike for the lack of consideration of capacity constraints when planning lots sizes. As one group of authors writes, "MRP systems in their basic form assume that there are no capacity constraints. That is, they perform 'infinite loading' in that any amount of production is presumed possible…" [19]

For some types of industries, like heavy manufacturing, this limitation is an annoying inconvenience. With finished items requiring high labor inputs, the primary capacity constraint is often availability of skilled workers to do the job. If high production levels press the capacity of available trained labor, more workers can be hired or existing workers can be retrained. In other situations, such as the process industries, lack of capacitated planning and scheduling is a much more serious matter.

The process industries are asset intensive with huge investments in long lead-time equipment. In this case, adding additional capacity is not a short-term managerial prerogative so it becomes imperative to get the greatest amount of capacity utilization possible through scheduling methods that find the optimal solution and consider dynamic capacity constraints. The lack of capacitated MRP is such a serious issue that some leading companies have declined to use MRP for planning and scheduling.[20] While the algorithms to do aspects of capacitated MRP (CMRP) are available, the drawback to implementation is partially dependent on lack of real-time data needed for a meaningful solution. To deal with dynamic demand for end items, manufacturers must account for capacity constraints at all levels of the supply chain. This ambitious goal remains elusive for most firms.

RFID technology overcomes one barrier to the implementation of advanced algorithms for capacitated MRP by providing a continuous stream of data for mathematical programming models to achieve CMRP in practice. Although there are a number of complicating factors that limit the widespread use of advanced models, a major drawback appears to be schedule stability.[21] Because of a lack of continuous data, replanning often occurs less frequently than needed. In addition, small changes inventory and production values caused by inaccurate counts or poor execution to plan (for both production and the sales forecast) also contribute to the schedule stability problem. The combination of these two factors can create large changes in out-front schedules and a great amount of instability within CMRP. Having a continuous stream of data allows quick adjustment to variances and frequent updates. If the proper buffers exist, a stable schedule results with only minor changes occurring over the time horizon with each new planning run.

There are several documented examples of the application of CMRP in industry.[22, 23] Most notable are early applications achieved in the semiconductor industry.[24] The approach uses large-scale linear programming (LP) to accomplish CMRP with the goal of improving on-time delivery. The authors note that before implementing the LP approach, sector-wide planning took place only once per month because of the poor quality and availability of data on demand, work in process and inventory. Essentially, planners always had incomplete information. A large part of the project included design of databases to feed the LP planning model and the development of standard ways to represent data. In the end, the authors state that data accuracy, availability, and timeliness were significant factors in the overall success of their efforts to implement CMRP as a management tool.

These are just a few examples of how RFID technology will change the nature of ERP systems in practice.

Reliable data capture along with software to manage EPC data in a timely manner are critical elements for success in creating the granular information needed for the supply chains of the future. In time, the EPCglobal Network and RFID will create an "Internet of things" that will have a significant impact on business. Harnessed with ERP systems, early projections call for RFID to deliver significant operational improvements, including: [25]

- Increased revenues of up to 1 percent from improved quality and customer service.
- Decreased Cost Of Goods Sold (COGS) of 1 to 5 percent from improved overall equipment effectiveness.
- Reduced working capital of 2 to 8 percent from reducing raw materials, work-in-process and finished goods inventories with shorter cycle times and better visibility.
- Reduced fixed assets of 1 to 5 percent from better maintenance and utilization of plant equipment.

To capitalize on this productivity trend, leading software companies such as SAP have already developed software modules to utilize data from RFID systems.[26] With time, this new source of data will become an integral part of ERP systems and decision-making in business.

CHAPTER 14

Building a Business Case for the EPCglobal Network

With the current state of the EPCglobal Network and RFID technology, it is feasible to do large scale tagging of objects within supply chains. This opens new possibilities for gaining real-time information, which line managers and executive's value as an important part of running a business.[1] By enabling real-time information, RFID technology has the potential to improve monitoring and coordination internal and external to the firm.

Given these capabilities, companies desire to identify areas where RFID technology can provide bottom line results in terms of reduced costs, better customer service, and improved profits. Building a business case for RFID, given current prices for tags, readers, and IT infrastructure, is a top priority that will determine the rate of future adoption.

However, as a practical matter the calculation of costs and returns on investment (ROI) becomes difficult because many elements of RFID technology fall into the category of corporate overhead. Application of tags to individual objects represents the only true variable cost. Yet even tag costs can change a great amount depending on the quantity purchased and the overall quality of the tags. Since the cost of tags should decrease within the next several years, the primary cost of RFID will result from changes to information technology (IT) infrastructure.

For most firms IT infrastructure is overhead that supports many different functions. Often it is hard to assign a proper allocation of overhead that is a fair representation of the amortized asset value for specific business processes. Further, it is also difficult to identify both quantitative and qualitative benefits that arise from RFID technology. With a bias toward high returns and quick paybacks on investments, there is a need to develop methods for fairly calculating the financial and qualitative impact of RFID technology in practice.

The overall costs of implementing RFID are substantial even by the standards of large corporations. Early projections are raising concern about the short and long-term ROI for the technology. One study shows that each of Wal-Mart's top 100 suppliers would have to spend between $13 and $23 million on RFID for the technology to be fully effective.[2] Another study conducted about the same time shows that of 24 companies independently surveyed, only one indicated it anticipates getting an acceptable return on investment in less than two years.[3] With these projections, it becomes extremely important for companies to analyze the ROI of RFID using proven tools established by industry leaders.

This chapter provides a case study and a method to evaluate the costs and benefits of RFID technology based on analysis conducted by the Dell strategic supply chain group.[4] The results of the study include an initial means to evaluate RFID technology that is applicable to other firms in other industries.

Before exploring ways to evaluate economic contributions, it is important to understand why the role of infrastructure is critical to the success of building an Internet of things. Infrastructure issues trace to the fundamental difference between EPCglobal Network and the traditional application of RFID.

RFID Infrastructure

As noted in Chapter 2, the majority of RFID applications have been proprietary. From an investment standpoint, RFID can offer acceptable financial returns for limited scope projects. In this situation, all investments associated with RFID can be easily identified.

An example is the application of active tags to railroad cars. This has been in place for more than ten years.[5] The cost of the tags and infrastructure to support these types of closed loop RFID applications are identifiable in that all computing systems are stand-alone. This makes the job of financial evaluation straightforward.

In contrast, interoperability between tags and readers characteristic of RFID technology is essential for wide-scale application within supply chains, but also complicates financial justification because the system IT infrastructure becomes a significant element of corporate overhead. This overhead might not contribute directly to the benefit of companies, but rather the customers served by companies. The rollout of RFID technology by Wal-Mart is an example where few manufacturers are receiving any initial benefit.[6] Since RFID is a supply chain wide technology, companies should expect that "much of the value of RFID will not be generated within the four walls of the warehouse or store, but instead, will depend on close cooperation between supply chain partners."[7]

This raises some interesting questions concerning the long-term trend of supply chain integration. Historically, integration has taken the form of tightly coupled relationships with suppliers characterized by greater information sharing, improved coordination, and joint performance measures.[8] This has led to the predominance of dyadic relationships that "rarely span across more than two adjacent partners in a supply chain."[9]

Though these tightly coupled relationships have produced results in terms of better coordination and reduced inventory,[10] there remain opportunities for multi-tier coordination to reduce the "redundancies in the supply chain that would otherwise not have been considered."[11] In this view, manufacturers, distributors, raw material suppliers, and retail outlets all work together to achieve supply chain efficiencies specifically in the areas of reduced inventory and improved customer service. Through supply chain coordination, it might be possible to reduce the impact of the bullwhip effect that often causes devastating swings in demand and inventory levels especially for suppliers located deep in the supply chain.[12]

Some have even gone as far as to predict that in the future companies will not compete directly, but rather, entire supply chains will compete against each other. The theory behind this prediction is that tightly formed supply chains closely resembling a vertically integrated company will generate much greater efficiencies and lower costs as compared to traditional business organization where independence is the norm.

While this is an interesting concept, the reality of the supply chain versus supply chain competition is not straightforward. It should not be assumed that this is a universal model of future organization.[13] Suppliers seldom conduct all of their business with a single customer or belong to a single supply chain guided by a channel captain.

However, beyond the questions of supply chain to supply chain industrial organization, RFID does provide the theoretical capability to gain detailed information, at the unique identification level, about mult-tiered supply chains. The most beneficial information obtained through RFID technology would be real-time inventory for finished goods, work in process, and raw materials at specific locations. Assuming that RFID can provide this information, there are several mathematical models that could be applied to optimize an entire supply chain spanning many individual firms.

Though this has value to companies, there remains the question of who will pay for a supply chain wide infrastructure that will provide detailed inventory information. Those that benefit from improved information flows might not be the same companies that must make the investments in the infrastructure needed to produce the information. This complex issue results from the new forms of supply chain visibility at the individual end-item level that RFID technology can create.

As a starting point in the analysis of RFID ROI, either inside a company or within an entire supply chain, there are only two ways a new technology can create benefits; reduction of cost or increasing sales.[14] According to a recent study, RFID has the potential "to reduce costs in five ways and increase sales in two." [15] Costs can be lowered through:

- Labor savings
- Reduction of theft
- Reduction of disputes with trading partners
- Reduction of excess inventory
- Reduction of spoilage/obsolescence

Sales can be increased through:
- Reduction of out-of-stocks
- Greater responsiveness to the customer

These categories of benefit creation could be individually applied anywhere within a supply chain, although there are few if any means of deciding how to divide global benefits among supply chain partners in proportion to infrastructure investment. The eventual wide-scale deployment of RFID will likely depend on working out the details of how to share benefits and investments in a multi-tiered situation.

Though there are few concrete examples to demonstrate financial evaluation techniques for RFID, some leading companies have undertaken early attempts to establish robust methods for financial analysis. In the absence of a universally accepted conceptual model to calculate RFID ROI, the case study approach provides the best short term means to gain insight about this difficult problem.

The next section provides a brief background summary of Dell Corporation that will set the stage for a case study discussion involving the calculation of the ROI for RFID based on specific business processes within the company.

The Dell Case

Starting with $1,000 in capitalization, [16] Dell Corporation has built a computer business that has achieved $50 billion in revenue within 21 years and has a present market capitalization of $100 billion.[17] This growth has occurred without acquisitions or mergers. An important aspect of this success has been something called the direct business model, originally put into practice by Dell starting in 1984.

By bypassing the dealer channel, Dell managed to sell directly to the customer and build computers to order. This eliminated the re-sellers' markup along with the costs and risks of carrying large inventories of finished goods. The direct business model has given Dell a substantial cost advantage and the ability to obtain valuable direct information from customers.[18]

In essence, the underlying key to the direct business model is Dell's insight on how to integrate different approaches such as customer focus, supplier partnerships, mass customization, and just-in-time manufacturing, into a unified whole. This insight enables "coordination across company boundaries to achieve new levels of efficiency and productivity, as well as extraordinary returns to investors." [19]

However, other attributes of the company beyond the direct business model also contribute to its success. The company has a cultural tradition that emphasizes day-to-day execution and consistency along with a strong bias toward action and decision-making based on data. There also exists a tradition of innovation in the details of daily operations. This emphasis has proven effective for Dell in terms of cost control and cash flow. In the words of the CEO Kevin Rollins, "we challenge our people to substitute ingenuity for investment." [20]

With this culture, new technologies such as RFID must pass a rigorous test to win acceptance within the company. "We're very risk averse," states Rollins.[21] He adds, "Occasionally our managers develop emotional connections to businesses that they really want to drive. But we make them prove the opportunity to us, and if we're not convinced, we don't move forward. We avoid areas where it's not clear we can be successful." [22] In this case, "proof" means a plausible financial analysis based on hard savings. Dell makes few decisions based on qualitative assessments or gut feel.

The RFID Scorecard

The Dell supply chain is unique in that it supports an assemble-to-order inventory strategy that entails high though-put volume and short lead times. Previously, most assemble to order inventory strategies involved low volume and long lead times. Typical examples included the automotive industry.[23]

Figure 14-1 shows a diagram of the Dell supply chain from beginning to end. Suppliers stock parts in logistics centers located close to Dell assembly plants. The plants assemble computers as needed to fulfill orders placed directly from customers. After assembly, the computers are shipped to the Dell Merge Center where other components, such as monitors or peripherals, are combined to form a complete order. This represents a merge in transit capability that is considered among the most advanced in American manufacturing.[24] Once all of the components are merged, the final product is shipped to customers.

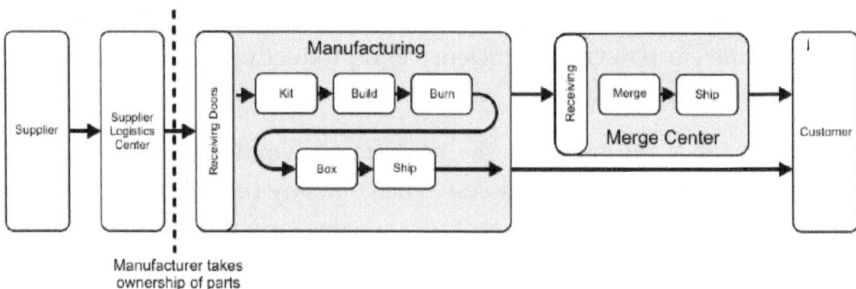

Figure 14-1 – Dell's Supply Chain – Overview

Taking insight from various efforts to evaluate corporate balance sheets and supply chain costs,[25] Dell has designed a scorecard approach for the financial analysis of RFID applications within its supply chain (See Figure 14-2).

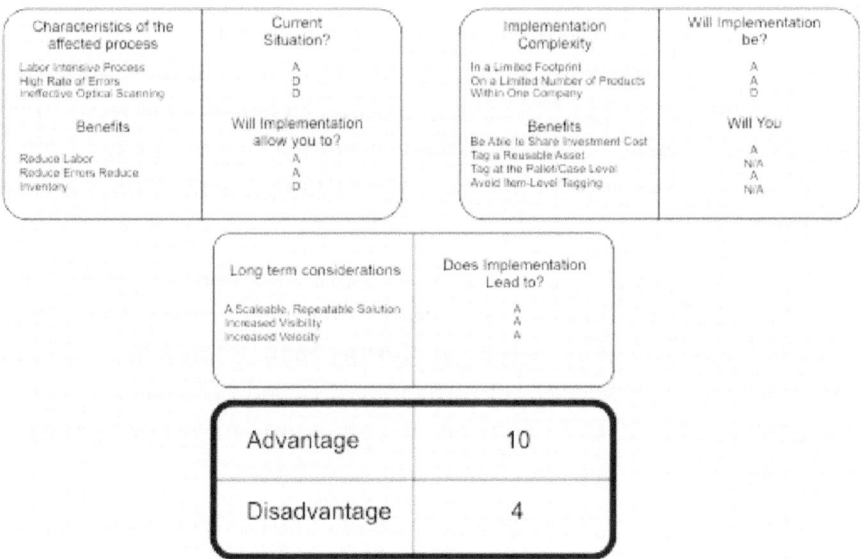

Figure 14-2 – Radio Frequency Identification Scorecard

Advantage (A) – The application of RFID is an advantage over an existing business process
Disadvantage (D) – The Application of RFID will not improve an existing business process

The scorecard is a simple set of critical questions designed to assess if a particular RFID application is worth pursuing. In the case of Dell, each question is weighted to reflect the importance and strategic direction that management wants to promote. After completion of the scorecard for each business process, all of the scorecards are gathered together and evaluated relative to each other. This provides a simple yet effective method to screen business processes for the best candidates. Once a subset of high potential candidates are identified, detailed financial analysis is conducted. Using this approach, Dell is able to screen a number of businesses quickly, focusing only on the best candidates for detailed analysis of hard savings.

Figure 14-2 shows an example for a high potential RFID application (Business Process A) at Dell. This simple analysis identifies the current state of a business process and the potential benefit if RFID were employed. The intent is to identify situations subject to high rates of error and that are labor intensive in terms of tracking and tracing. Applying RFID in these situations will give Dell a higher payback.

The example detailed in Figure 14-2 shows a situation where it makes sense to proceed with detailed analysis of RFID. An advantage score of 10 and disadvantage score of 4 indicates application potential. Dell has applied the scorecard to the following business processes: tracking totes and trays within plant operations (assembly); tracking notebook computers from manufacture in Asia to delivery in the US; and tracking the inbound delivery of microprocessors to domestic manufacturing plants; other business processes.

In most of these cases, the score is 8 advantage, 5 disadvantage. All business processes are evaluated and ranked using the scorecard. Those processes that rank highest are selected for further analysis. There is no specific cut-off for processes selected for further analysis and processes considered inappropriate for RFID. In this way, the scorecard is a relative means of evaluation.

In addition, Dell uses the scorecard to identify areas where existing data-capture technologies such as barcodes are not working to peak performance. Since bar codes represent a means of ubiquitous identification, it is likely that RFID technology, which overcomes some of the limitations of bar codes, might prove a better alternative.

One common theme from all early RFID implementation efforts is that achieving an acceptable return is difficult when application occurs on a limited scale. Being a networked based technology, there is no question that the full benefit of RFID will not be achieved until all firms within a supply chain implement the technology. In this regard, implementation resembles that of a ground based telephone system. A partial network of telephone lines does improve communication; however, it is only through coast-to-coast wiring of every home that the full potential of a telephone network can be realized.

The emphasis at Dell has been to identify opportunities where acceptable returns can be achieved through limited application of RFID technology. This assumes that much larger benefits will probably happen through full implementation; however, a limited project offers the opportunity to become familiar with RFID technology while still achieving positive financial results in practice. Smaller projects also mean less risk. In many ways, this approach resembles experimentation needed for all innovations, with each experiment being chosen based on the likelihood of financial success.

Dell examines RFID justification as it does for all new types of technology. Hard savings take precedence in making investments with particular emphasis on savings in reduced labor, fewer errors, and inventory carrying cost. Though RFID technology has great potential to make significant contributions in all three of these areas, partial implementation increases the difficulty in justifying leading edge applications that in the long-run will return the greatest amount of value to Dell.

A Conservative Approach

To reduce the cost of the initial implementation, Dell has taken the approach of looking for a promising subset of the supply chain for early applications within their own operations. By narrowing the scope of application, less hardware such as readers for tags is needed and there are fewer coordination problems.

In addition, it is not possible to wire an entire distribution center or factory as a starting point for RFID. Rather, Dell looks for a defined location within a facility and specific individual product flows. This enables the test to be completed with the minimum of tags and readers, thus reducing the initial hardware investment. Using this approach also diminishes the impact to ongoing operations. However, care must be taken that this does not over-simplify the issues relating to RFID applications between trading partners.

As a final comment, all of the scenarios examined by Dell involve tagging at the case and pallet levels. To date, there has been no analysis of tagging finished goods shipped to customers such as home or corporate users. For the realistic future, Dell will focus on RFID applications that deal with supply chain issues that include suppliers and internal plant operations rather than customer applications that might involve computers or printers.

Building the Business Case

Once the scorecards for each process identify the best candidates for RFID technology, the next step is to do the financial analysis of the benefits and costs. At Dell, all benefits must come from hard savings including reduced labor, fewer errors, and lower inventory carrying cost.

There are many important questions to ask at this stage. How many readers are needed? What are the incremental computing requirements? How are tags applied? What software will manage the data provided through RFID?

Dell has concluded that although industries have focused on the price of tags as the biggest hurdle, the largest cost is in systems integration. Figure 14-3 shows a *mock* payback calculation for a particular business process. In this business case, the payback was about one year.

Benefits		
Reduce Labor	80	
Reduce Errors	75	
Reduce Inventory	0	
Total Yearly Benefit	**$155**	
One-Time Costs		
Hardware		
Readers	5	
Application Servers	8	
Data Storage	4	
Software		
Operating System	2	
RFID and Database Software	18	
Subtotal for Hardware and Software	$37	
Installation and Integration Services	$50	
Total One-Time Costs	**$87**	
Recurring Costs		
Support and Maintenance (15% of Hardware and Software costs)	$5	
Number of Cases and Pallets Per Year	100	
Cost Per Tag	$0.25	
Annual Tag Costs	$25	
Total Yearly Recurring Cost	**$30**	
Payback Calculation		
Yearly Return @ Stabilization (Annual Benefits Less Recurring Costs)	$125	
Installation, Integration, and Stabilization Time (Years)	0.3	years
Years to Recoup One-Time Cost (One-Time Costs/Yearly Return)	0.7	years
Payback	**1.0**	years

Recap
Yearly Return 155 − 30 = 125
Years to Recoup One-Time Cost 87/125 = 0.7 years

Figure 14-3 – Sample RFID Business Case

Making the Decision

After completing the scorecards and business case analysis, Dell has a structured approach for going forward that involves three options.

Go

The scorecard shows an advantage and the business case has an acceptable payback period or ROI based on capital hurdle rates. In this instance, the project is implemented immediately.

Stop

The scorecard shows no advantage over existing processes in the application of RFID technology. The project is stopped.

Hold

In the situation where the scorecard shows an advantage but the business case does not quite show returns that meet corporate objectives, the project is put on hold pending further developments. As costs change, it might become feasible to go forward with the project. For example, the mandates for RFID technology from Wal-Mart, the Department of Defense and the Food and Drug Administration will drive greater production of tags, readers, software, and systems integration. The increased volume of activity will result in economies of scale and more intense competition among vendors. In addition, technology performance will improve over time.

For most situations at Dell, RFID technology currently falls into the category of hold. This is the case because the calculations for justification depend entirely on hard savings. Since Dell has already invested billions of dollars to develop business processes that are state of the art, especially in the area of minimizing inventory, it is often hard to find overwhelming savings from RFID that justifies immediate implementation. However, this could all change in a relatively short period as the costs of RFID technology decrease.

The greater value of RFID technology may be in the realm of customer service. It is very difficult to measure these benefits directly, however every business knows that when done right customer service is a factor in long-term sales growth. Being able to track and trace parts by serial number, calculate the reliability of critical components such as hard drives, and deliver service by treating each computer sale as a unique event offers great benefit to customers. This type of capability also offers differentiation from competitors who have not yet developed methods to treat each customer as a unique entity.

Conclusion

As a broad statement, there is concern in industry that RFID mandates by Wal-Mart, the Food and Drug Administration, and the Department of Defense are the only driving forces for industry to adopt RFID technology.[26] Many believe that an overriding business case does not exist for the immediate implementation of RFID in a wide range of industries because of general uncertainty in terms of ROI and functionality, and because there are so many alternative IT and operational investments that compete with RFID.[27] However, as costs decrease and tag read rates improve, this situation could turnaround very quickly.

Of particular importance to the ROI and initial investment in infrastructure for tags and readers is the strategy for implementation.[28] Retailers desire to deploy tags and readers geographically targeting select distribution centers. This regional approach helps to control the cost of deployment in terms of infrastructure. However, it also means that all products handled in a particular distribution center must be tagged to obtain the benefits of RFID. This includes slower turning items where the advantage of RFID is marginal.

On the other hand, manufacturers want to deploy on a select product category basis with national distribution. In this case, manufacturers see advantages in improving in-store stock rates on fast moving items and reduced theft and counterfeit for high value items.

These opposing views of deployment held individually by retailers and manufacturers will need to be reconciled before substantial progress can occur in situations where RFID is mandated.

CHAPTER 15

Enhancing Revenue Using the EPC[a]

Most of the proposed applications for the EPCglobal Network and RFID technology have focused on innovative ways to reduce cost within the supply chain through tagging cases and pallets. While this cost orientation should create significant savings given continued improvements in tag costs and read rates, there is also the possibility of creating value in other ways. Within the consumer goods industry, the data generated by the EPC could be an effective tool to enhance revenue. Achieving top line growth is the goal of every firm and an area of great interest to senior management. Driving revenue growth might prove to be the ultimate value of the EPC.

There are three basic ways to grow revenue.[1] First, a company can stimulate sales of existing products through improved pricing, promotion, and advertising. Second, a firm can increase overall sales though introduction of new products. Finally, growth can occur though acquisition of businesses lines or entire companies.

Given the competitive nature of many markets, growth through selling more of existing products is becoming increasingly difficult and often causes a drain on profits because of rising advertising and promotion costs. As an alternative, growth through acquisition presents significant risks. Most corporate mergers do not achieve pre-merger growth targets.[2] Because of these business realities, many firms are looking to increase revenue through the introduction of new products. This is especially the case in the consumer goods industry, although new product introductions also bear great risk.

Many managers and business consultants feel that increasing the efficiency of advertising and promotion for existing products, along with improving the process for introducing new products into markets, is the most consis-

---

[a] The authors wish to thank Professor John R. Williams and Ching-Huei Tsou, both of MIT Auto-ID Labs, for technical guidance concerning this chapter.

tent overall strategy for achieving long-term revenue growth. Historically, the bar code has played a role in supporting revenue growth by providing product data on the volume sold through retailers for both durable and non-durable consumer goods.[3] This data has been instrumental in creating an entire family of market mix models that measure customer response to promotion and advertising, allowing retail managers and manufacturers to make better trade-offs in practice.[4] However, even with the large amounts of data generated from barcodes the science of marketing remains a "fact-based art."[5]

The EPC has the potential to go a step further by helping companies not only to understand what stores are selling, but also what individual customers are buying.[6] With the capability of unique identification for an item, the possibility exists of tracking customer sales at a level of detail not currently possible. This could lead to important insights concerning the demographic, economic, and geographic forces influencing purchasing decisions. At least one marketing services company has already managed to take the first steps in building customer profiles using transactional data gathered from barcodes.[7] The EPC will enhance this process by providing unique identification of individual customers.

Though it will be several years before tagging of all items in retail stores becomes economically feasible, situations currently exist where tagging single, high value items might make sense. The top candidates are durable consumer items such as appliances and electronics. These consumer items share a common characteristic in that the logistical shipping unit, a case, contains a single end-item. In these situations, tagging individual cases and the use of the EPC becomes a favorable option given the additional supply chain and marketing information that could result.

There are also opportunities to use the EPC on loyalty cards, a means of customer sales tracking used by many retailers such as CVS/pharmacy.[8] This type of application provides opportunities to do interactive marketing within a store, expanding the ways to send advertising and promotional messages to targeted individual customers.

In all of these cases, the use of the EPC to obtain detailed information on individual customer purchasing behavior presents significant issues of privacy and data security that retailers and manufacturers must address before widespread application can take place. Though privacy is an unanswered issue, the value of unique identification poses many yet to be realized advantages for customers, retailers, and manufacturers.

It is seldom the case that technological innovations like the EPCglobal Network and RFID technology take place within a vacuum. Even as manufacturers and retailers analyze its implications, there are other structural changes taking place in the consumer goods industry that could have an impact on the eventual use of the EPC. This is especially true in the area of advertising and interactive marketing.

The traditional notion of mass marketing and the concept of the economies of scale for advertising are changing rapidly. With new possibilities for reaching individual consumers through the Internet and other communication networks, advertisers no longer "treat consumers as homogeneous masses." [9] In this environment unique identification, whether through the EPC, a personal computer, or a cell phone, takes on new meaning and value to business. The next section discusses several emerging applications poised to change the way advertisers reach customers, and the way industry and the public view technologies such as the EPC.

New Developments in Advertising

In practice, only about half of all advertising is successful.[10] With the advent of commercial skipping devices like TiVo, along with the fragmentation of viewers caused by cable television, video games, and the Internet, the traditional business practice of mass communication through television and radio commercials is coming under increasing pressure.[11] Though this approach has worked well since the beginning of radio in the 1920s and television in the 1950s, the impact on today's audience appears to be diminishing. Companies well known for popular network television commercials such as the GAP, inc. and Anheuser-Busch are shifting advertising dollars to direct mail catalogs, cable TV, and the Internet.[12, 13] The primary reason given for these shifts in advertising budgets is recognition of the changing nature of viewership associated with network television.

As the growth rate of network television advertising declines, Internet-based advertising continues to grow rapidly. For example, the home pages of Yahoo, America On Line (AOL), and Microsoft's MSN are sold out for display ads "months in advance." [14] In some cases, online advertisers are seeing double digit increases in ad prices.

Even though online advertising totaled only 3.7% of the total ad dollars in the US during 2004, [15] ad response rates are comparatively high at about 3% for popular web sites such as AOL.[16] Some believe this high response rate is a result of the community aspects of a web site like AOL, where news, entertainment, and the opportunity to interact with other online users keep visitors returning and spending money.[17] As businesses such as AOL, Yahoo, and MSN refine the techniques of building a community, online ads should increase in importance.

With more Internet users responding to online ads, it becomes possible to measure a particular advertisement's effectiveness using such methods as econometrics to estimate how various factors affect sales.[18] Through various means, the Internet can provide a wealth of data needed for econometric models. As the online advertisement business matures, some companies such as Yahoo are hiring firms to track the effectiveness of search related ads. Concerning this trend, one author comments that "the effectiveness of search-related ads is seen as easy to measure – advertisers only pay for the ads' placement when people click on them, and can track when clicks translate into purchases." [19] This represents a major improvement in targeting a marketing message.

Marketing Innovations

Along with the new developments in online advertising, a number of innovative marketing ideas are emerging as retailers and manufacturers explore new ways to use technology. Three interesting ideas involve using wireless communication as a means to reward customers who visit a restaurant, employing an EPC loyalty card for interactive marketing within a store, and the use of cell phones as a promotional tool and to measure advertising effectiveness.

Walt Disney Corporation has filed a patent application for a system that would provide a portable media player to those who visit particular restaurants and purchase a meal.[20] The media player is capable of holding a Disney movie, music, games, or photographs and is WiFi enabled. According to the pending patent, users would receive an electronic code "that authorizes a partial download of a movie, video or other media file, which can be downloaded while in the restaurant." [21] With each subsequent visit to the restaurant, patrons would earn points and could download additional files.

This produces a powerful incentive to return to the restaurant and purchase more meals.

In another example, retailers and technology companies are experimenting with different methods of in-store interactive marketing. For some time, researchers have known that a significant relationship exists between a store's marketing environment and customer purchasing behavior.[22] Reaching the customer with a promotional message at the instant of purchase is a powerful method of increasing sales and brand loyalty. The only barrier holding retailers and manufacturers from doing more interactive marketing is the expense and lack of technology at the store level.

However, with the imminent changes in the mass communication industry as advertisers attempt to reach an increasingly fragmented audience, retailers and manufacturers view in-store interactive advertising as the wave of the future. Even established consumer goods manufacturers like Proctor & Gamble, the company that popularized mass market advertising almost one century ago, are shifting to a strategy of pitching brands directly to customers as they shop in stores.[23]

To build on this trend, one company has created an entire business based on in-store, interactive digital signage that responds in different ways to customer traffic along with methods to evaluate the effectiveness of the posted message.[24] The fundamental technology employs video camera surveillance. Most of these in-store applications are outside of the United States. Anecdotal accounts put product sales gains at about 10% through the use of digital, interactive signage.[25]

In another situation, IBM has proposed technology that uses an EPC embedded in clothing displayed for sale. As a potential customer selects an article of clothing from a rack or shelf, a reader scans the tag and a digital display located close-by shows information about potential accessories or other types of clothing that match.[26] This type of technology reduces search time inside stores, satisfies the customer and increases the probability of a sale. In a similar way, loyalty cards might contain an EPC, allowing customized messages to be displayed as customers walk through a store.

Interactive marketing and monitoring provides other benefits indirectly linked to increasing sales. One company has developed a means of using RFID technology to monitor the timing of point-of-sale (POS) displays sent by manufacturers to retailers for set-up in stores. Goliath, Inc. reports that about 50% of the time, stores "do not deploy the displays in a timely and consistent manner." [27] The company provides an in-store monitoring service for consumer goods companies to audit the timeliness of POS displays. By combining this information with check-out scanner data gathered from barcodes, a true picture of the impact of POS displays on sales emerges. In this case, data from RFID supplements existing data obtained from bar codes.

A final example involves an interesting situation in Japan where cell phones are used in an elaborate system of interactive advertising. As part of a promotional campaign, Northwest Airlines has constructed billboard advertisements in Tokyo that also contain two-dimensional bar codes. A passerby can scan the bar code posted on the billboard with a cell phone specially equipped to decode the information contained in the bar code. Once decoded, the message automatically directs the cell phones' web browser "to coupons, games, or further details on a product" located on the Internet.[28] There are 30 million cell phones in Japan equipped with QR technology, the means needed for scanning and decoding of large bar codes located on billboards. The technology-oriented Japanese find QR scanning a curious pastime. Advertisers find QR technology a valuable source of information about traffic patterns in cities and the advertising effectiveness of billboards.

In all of these cases, unique identification plays an important role in enabling the interaction between consumers, retailers, advertising and marketing research companies, and manufacturers. As business moves into an age of mass customization for consumer products, markets will continue to fragment. Given this ongoing trend, the EPC should take on greater importance as a basic element in dealing with the complexity of future markets where product life cycles are shorter and new product introductions are frequent.

Beyond enhancements to advertising and promotion, the next section explores ways the EPC might improve the logistics of new product introductions. This builds on another important trend in business, the integration of marketing science, engineering technology, and supply chain management.

New Product Launches

The launch of a new product is a challenging task that every consumer goods firm must face. With individual stores now carrying up to 100,000 different items,[29] the ongoing reality of marketing is that the average customer encounters over 1 million different stock keeping units (SKUs) across all channels of distribution. Yet the typical family gets 85% of their needs from only 150 SKUs.[30]

Complicating matters, each year there are over 10,000 new product introductions in the non-durable consumer goods segment alone.[31] About 80% of these new introductions are food products. Of the non-food items introduced each year, about 80% are health and beauty aids.

Few reliable estimates exist concerning the number of new durable consumer goods introduced each year. Counting all of the line extensions within individual brands, and the wide range of durable consumer products ranging from personal computers to lawn equipment, the number must be staggering.

Given these conditions, manufacturers and retailers need to focus on innovative ways to advertise and track new products and to target potential customers. Future success will depend on using technologies that create opportunities for detailed analysis, and the ability to optimize advertisement expenditures during critical new product rollouts. With slotting allowances charged by retailers accounting for 16% of product introduction costs, it becomes critical that sales take-off for each retail outlet to realize a positive return on this investment.[32]

One of the most important factors influencing new product rollouts involves the geographic forces that affect adoption by individual customers. Commonly called spatial diffusion, this area studies the rate and pattern of adoption for a geographical area based on the frequency and type of advertising, demographics, and distance to retail outlets, along with other elements of the market mix such as pricing, promotion, and tactical product positioning versus competitors. Besides the general goal of describing and understanding customer behavior, the study of spatial diffusion also seeks to build mathematical models of the adoption process through time. This model building approach has practical value for retailers and manufacturers in providing general guidelines about how consumers adopt new products. Future trends in information technology, discussed in the final chapter of this book, will allow these models to exist in a network with the prospect of rapid linkage to data for real-time analysis.[33]

Marketing managers and academic researchers realize the importance of studying spatial diffusion. A research article on the subject notes, "Managers who understand the geography of the processes by which consumers change their behaviors can be much more successful in launching new initiatives and can make much better use of their resources while doing so."[34]

For all types of consumer goods companies, the initial spatial pattern of adoption is an early indicator of longer-term success or failure. This becomes critical information for the process of establishing a new national brand. Achieving rapid market penetration in select areas of the US is vital in gaining profitability and forestalling competitor response, especially since national rollouts are costly, characterized by historical risk of failure, and seldom done in practice.[35]

While the development of marketing research in the US economy has been a fundamental reason for the growth and sophistication of the consumer goods industry, the introduction of new products into select markets continues to represent an area of great inefficiency in terms of supply chain and advertisement costs. For example, in examining the entire new product introduction process from initial conceptualization to successful launch, only 1 in 58 are successful in the food industry.[36]

Assuming a new product is properly developed and is reasonably matched to customer needs, [37, 38] a significant reason for failure lies in the amount, quality, level of aggregation, and timeliness of data available to managers concerning the rate of geographic customer adoption. With a lack of data, marketing managers have no chance to take a rational and analytic approach to decision-making about advertising and supply chain factors such as level of customer service that influence the degree of success for a new product.

In a typical product launch, the only sources of near real-time data are point estimates of aggregate demand from retailers obtained from scan data, consumer feedback from various types of interviews, namely focus groups, and household market research sampling by companies such as IRI and Neilsen. Often data from these sources give conflicting views of new product success.[39] In addition, none of these sources include the detailed geographical data needed for observation and modeling of spatial diffusion. Sometimes important geographical links do exist in the data although these links must be culled from daily purchase transactions. Technologies such as the EPCglobal Network and RFID offer the future prospect of obtaining real-time geographical data on spatial diffusion within a market, serving as an important data input to decision-making.

Further, there currently is a lack of efficient ways to visualize and organize the data needed to make spatial diffusion a management tool suitable for daily decision-making. There are no open computing systems capable of matching mathematical models describing spatial diffusion to geographic location data obtained from such sources as loyalty cards, and other future means that might include virtual private networks.[40] While history-based methods such as share and distribution analysis are valuable market research tools, spatial diffusion through its focus on the dynamics of customer behavior and real-time analysis of data offers much more potential to gain insight into the success or failure of a new product. This is especially the case for high tech products such as consumer electronics.

Recent advances in three-dimensional geographic visualization of urban areas and improved ability to manage different types of digital maps through browsers open a range of possibilities for the widespread application of spatial diffusion modeling. The next two sections analyze the details and value of spatial diffusion along with a more detailed discussion of data. The remaining parts of this chapter examine several examples of how information technology can advance the practice of using spatial diffusion as a marketing decision-making tool.

Spatial Diffusion in Markets

Regardless of advertising on television, radio, magazines, newspapers, and the Internet, word-of-mouth remains a powerful force in marketing. Some initial survey data suggest that recommendations from a "friend, expert, or relative" influences up to 80% of all purchases.[41] Large and small companies alike hold this type of advertising in high esteem.[42]

Geographical proximity is perhaps one of the strongest determinants of the rate of word-of-mouth communication. In a first attempt to understand the

impact of product recommendations made through interpersonal communication, a study conducted in the 1940s focused on the diffusion of technological innovation within an agricultural setting.[43] The authors concluded
that personal networks, in conjunction with mass communication, played
an important role in the decision by each farmer to purchase hybrid seed
corn. This work also categorized adopters into types and developed curves
of cumulative market share through time.

During the late 1960s, Professor Frank M. Bass took this work a giant step
further by developing a model that predicts the number of users who will
adopt a new product.[44] This model has been applied to describe the "sales
of televisions, clothes, dryers, dishwashers, refrigerators, and other consumer durables." [45] Through time, researchers have added modifications to
the basic model and applied it to a variety of products and situations. An
interesting example is the diffusion of a free email service (Hotmail).[46] In
this case, the Bass model made an accurate prediction of adoption rate
when matched to historical data.

Through the intensive examination of new product introductions, Professor
Gerard Tellis goes even further in developing a large set of aggregate data
and modeling techniques to determine the time of product take-off in the
United States and other countries based on different cultural and categori-
cal variables.[47, 48] This is valuable to know given that executives like General Electric's Jeff Immelt predict that as much as 60% of the companies'
growth during the next decade will come from overseas, specifically from
developing countries.[49]

The basic diffusion models made famous by Bass and others provide
insight and an analytical tool for marketers to make decisions. The next
stage in the development of this type of analysis involves the introduction
of spatial data. For many years, economists have used spatial data to predict the location of industry, the expansion of cities, and the clustering of
similar businesses within regions.[50] Making the transition to using spatial
data as a common element of marketing and supply chain analysis requires
the capability of unique identification that the EPCglobal Network and
RFID technology can provide. With this capability, it becomes possible to
analyze the geographical forces that influence word-of-mouth communication and the adoption of new products.

The modern history of spatial diffusion in marketing begins with research conducted by William H. Whyte in 1954.[51] In this study, aerial photographs of urban Philadelphia were used to identify homes that had window mounted air conditioning units, an innovative home appliance that gained popularity in the immediate post WWII era.

After compiling the data, Whyte observed that city blocks of equal size and demographics had sharply different rates of market penetration. For example, some city blocks had 18 air conditioners while others, located in the same general area, had only three. This study provided the first evidence that adoption of a new product, like an air conditioner, occurs in clusters and is not homogeneous over a geographical area.

Three Important Stages

Since the Whyte study, there has been a great deal of research about spatial diffusion in the context of geography, sociology, urban planning, and environmental science. However, little research has related to the introduction of new products and services to consumers in a defined geographical area. In theory, spatial diffusion for consumer products comprises three relatively predictable stages: 1) lead adopters, 2) neighborhood effects, and 3) consolidation of adoption.[52] As a generalization, the outcome of the three stages is an S-shaped curve of cumulative adoptions through time. The main driver of each stage is the rate of information dissemination to consumers within a market. Temporal aspects of spatial diffusion have particular importance in analyzing dynamic customer behavior, an under researched area of marketing.

When a new product is introduced into a market, no information about it exists. Leading adopters piqued by curiosity will likely be the first to make a purchase. The geographical pattern of these lead adopters is random. Since there are relatively few lead adopters in any market, the random pattern is characterized by significant physical distance between purchasers.

If experience with the product is positive, there tends to be local word-of-mouth communication by lead adopters, which influences others within close proximity to consider purchasing the product. In addition, advertisement often provides an impersonal means of communicating information about a new product across a wide geographical area. Because of both advertising and word-of-mouth from lead adopters, there tends to develop a neighborhood effect where clusters of adopters begin to form within a geographical area. These clusters continue to consolidate as more information about the product is diffused within the neighborhood network.

Finally, the overall market reaches a saturation of information about the new product. Lag adopters, who are inherently cautious in taking risks, begin to make purchases. Clusters of adopters tend to merge forming a pattern of complete market penetration.

Figure 15- 1 shows a classic case of spatial diffusion for a successful new product introduction through time. Figure 15-2 shows a different situation where spatial diffusion never advances from state 1 to stage 2. The result is a failure of the new product introduction.

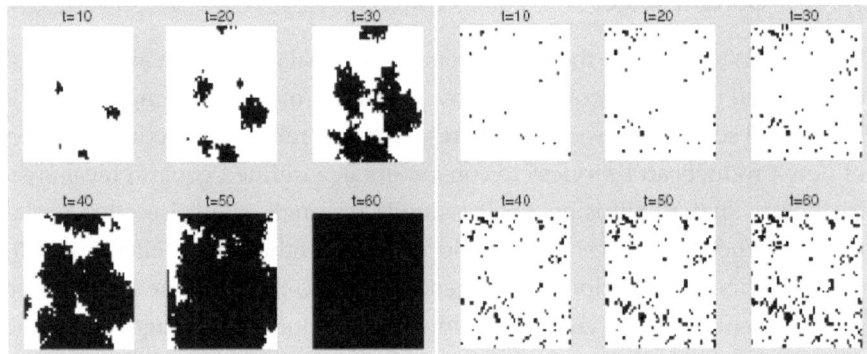

Figure 15-1 – Successful Introduction   **Figure 15-2** – Unsuccessful Introduction

Source – (Garber et al. 2004)

Used with permission, Marketing Science – INFORMS

Implications for Marketing and Supply Chain Managers

The characteristics of spatial diffusion raise a number of implications for managers. First, the rate of diffusion is very important in determining the overall success of a new product. Keeping in mind that lead adopters tend to be randomly distributed within a geographical area, understanding their spatial pattern gives some insight into the future rate of diffusion. This is important because the formation of clusters (stage 2) is partially rooted in communication between lead adopters and others in close proximity.

Any management effort designed to increase the rate of diffusion from stage 1 to stage 2 reduces the chances that competitors will dislodge a new product introduction. Once consolidation takes place in stage 3, competitors must confront the cost of getting a customer to switch in order to increase market penetration. This presents a barrier to gains in competitor market share.

Rate of diffusion also has another important implication. If a slow pattern emerges, as indicated in Figure 15-2, this might be an advanced indication of new product introduction failure. If lead adopters are dissatisfied, neighborhood effects will not take place and clusters of adopters characterized by stage 2, will never properly form. Without this base, advance to stage 3 saturation becomes nearly impossible.

Some advocate using spatial diffusion for predicting the overall success of a new product during the initial stages of the launch. An interesting research article develops indicators that can quickly show management the probability of success, and highlight opportunities to change marketing strategies such as advertisement, in-store promotion, and pricing to increase the rate of diffusion for a new product that is lagging.[53]

This work has important implications regarding demand forecasting for a new product introduction into a specific market. Time series forecasting methods frequently used in retailing and consumer goods manufacturing have little connection to predicting the spatial diffusion process because growth in demand is non-linear and dependent on external variables such as advertising. As a result, common forecasting techniques used in practice are ineffective when determining future demand for new products through time.

Improvements in new product demand forecasting have important implications for optimizing supply chain costs and customer service. If a new product introduction is successful, spatial diffusion will move quickly from stage 1 to stage 2 and demand will increase sharply for a specific market. If supply chain managers are unprepared to handle this increase because of an inaccurate forecast, significant stockouts at the retail level will occur. As previously mentioned in Chapter 11, an important study found that nearly 23% of consumers leave a store immediately in response to an out-of-stock.[54] This is devastating to new product success. Many new products have not progressed from stage 1 to stage 2 solely because poor customer service caused discontent in the mind of the consumer.

A final application of spatial diffusion involves optimization of spending on advertisement. Marketers have a number of mass media options to deliver a message to customers. Determining the optimal pattern and level of spending on advertisement represents a challenging problem in optimization. Because of the complexity of the problem, and the difficulty in obtaining real-time spatial data, historical demand trends often become the basic means to make media purchasing decisions. This represents an aggregation approach that simplifies the problem.

However, point estimates of demand commonly used in share and distribution analysis give few indications concerning the stage of spatial diffusion for a particular marketing area. This uncertainty leads to sub-optimal decisions concerning advertisement spending.

For example, a common opportunity for improvement involves the rate of advertising spending once a market has reached stage 2 in the spatial diffusion process. Some believe that further advertising upon reaching stage 2 accounts for lower marginal gains in sales at best.[55] By the time a market reaches stage 2, there is enough adoption, cluster formation, and momentum that further advertising has much less effect on overall sales.

Reducing media spending, just as overall sales for a marketing area are beginning to accelerate, appears counterintuitive if marketing managers have no understanding of spatial diffusion, or real-time geographical data to measure patterns of geographical adoption. Spatial diffusion offers deeper insights that allow marketing managers to make better decisions.

The Data Issue

To initiate spatial diffusion studies, data for new product sales must contain some aspect of time and location for a specific geographical area. Since the late 1980s, there has been a consistent effort by marketing research firms to gather "single source" data at the household level.[56] The EPC will play a future role in this type of data capture, both for marketing research studies and for new product sales to the public.

Single source data contains information on several independent variables and allows for causal analysis at the point of effect, either the store or the household level. Independent variables might include the impact of promotions, advertising, coupons, or the local competitive dynamics between stores. The level of detail is often on an individual product basis.

Obtaining large volumes of household level single-source data has been a historical limitation in conducting spatial diffusion studies. Improved technology is creating new opportunities to overcome this limitation through high tech market research techniques where individual household occupants agree to independent monitoring of their shopping, reading, and television habits.[57] However, this still only represents a sample of all households within a geographical area. At best, various in-home technologies might allow monitoring of perhaps several hundred thousand households across 24 countries including all marketing channels.[58] In some cases, the sample is not large enough within specific demographic groups for reliable spatial diffusion analysis, especially when considering large, developed economies like the United States where a great deal of diversity exists and population densities differ by a wide range.

At the store level, the situation is no better. In general, data gathered from barcode scanners provides little value in making short-term decisions about product strategy, advertising, forecasting, and spatial diffusion. This is because important demographic information and knowledge of other independent variables that might affect the purchasing habits of specific consumer segments are missing.[59] An additional limitation, important to the study of spatial diffusion, involves the lack of customer geographic information.

Some firms, most notably Target, have gone as far as to issue a branded Visa credit card in an attempt to get improved data about customer demographics and sales patterns for products purchased in their stores and elsewhere.[60] This follows a general trend toward using credit card information and other means as indicators of aggregate sales increases by category during the holiday shopping season.[61, 62]

Because none of these approaches provides a truly representative sample of total commerce, there is often divergence in the sales increases predicted, even between various credit card reporting groups. No indications exist that transactional data from credit cards is a fundamental part of spatial diffusion studies even though this data contains customer location and products purchased.

In one of the most practical and comprehensive efforts to date, a research study used data obtained from loyalty card information.[63] The purpose of this study was to understand the rate of spatial diffusion for loyalty card adoption; hence, the address for each adopter and the time of enrollment became the data needed to conduct the study. Advanced information technologies, discussed in the next section, can convert street addresses appearing in databases to earth coordinates, longitude and latitude, for automatic use in digital mapping. This is a powerful tool to visualize spatial diffusion through time.

With the loyalty card example, each enrollment results in a data element consisting of time and location within a geographical area. Loyalty cards also allow the ability to record individual customer transactions, potentially giving data concerning a new product's sales history. However, since those customers that hold a loyalty card represent a subset of all customers for a particular retail outlet and geographical area, the data obtained from loyalty cards is subject to sampling statistics. Since lag adopters are likely to hold a loyalty card in fewer numbers, no direct inference can be drawn unless a high percentage of all customers visiting the store hold loyalty cards and use them regularly to make purchases.

The growing trend toward online sales offers additional prospects for gaining the data needed to conduct spatial diffusion studies. When a customer purchases a product, it is shipped to a specific location at a specific time. Since online retailers capture all of this information, detailed geographical, time, and location data exist.

However, spatial diffusion studies might be of lesser value to online retailers, because the means of promotion and advertisement mostly occurs through the Internet, a medium that has no geographic boundaries. Although spontaneous, interpersonal word-of-mouth advertising remains an important element of marketing, product recommendations obtained though the Internet are taking on greater importance as millions of Blogs provide useful information to consumers and emails alert of pending new product introductions. In this instance, diffusion studies with no specific geographical component, like those first initiated by Bass, might provide greater insight concerning Internet communication and its impact on new product sales.

In all cases mentioned above, the central theme is the lack of unique identification associated with individual product at time of purchase. The EPC has a role in providing this information for high value, durable consumer products such as electronics and home appliances sold through traditional retailers where the purchase amount is generally one unit.

Perhaps the more important near-term role of the EPC involves increasing the efficiency of household market research studies and placement on loyalty cards as a means of increasing ease of use. This opens a number of possibilities for interactive marketing, understanding the true impact of advertising and promotion, and local spatial diffusion studies of new products.

While the data gathering capabilities of the EPC offer a number of beneficial possibilities for marketing analysis, there also exists the important issue of privacy. Obtaining detailed information about personal purchasing patterns is something that makes many consumer groups and the American public feel uncomfortable. There is always the possibility that personal data could be compromised through computer hacking.

As a case in point, the drug store retailer CVS/pharmacy has received criticism for making available to customers the purchase data from loyalty cards through a password secured web site.[64] The password security did not prove to be robust and this web site was hacked, allowing access to a history of recent purchases for each customer. Since CVS/pharmacy has issued 50 million loyalty cards, a great deal of private information existed on the web site for a large cross-section of the American public. Since the incident, the company has improved its security procedures in an effort to eliminate the future possibility of a data compromise.

Assuming that privacy and security issues can be overcome,[65] the integration of EPC and other data into digital maps holds great promise for building real-time systems that highlight the important aspects of spatial diffusion of new products. The final section of this chapter introduces digital mapping capabilities and data handling approaches needed to make these systems a useful management tool.

Visualizing Spatial Diffusion

Two powerful information-processing systems exist in the world, the human mind and the digital computer. The main connection between the two systems is the display of data and information, commonly called a user interface.[66] Maps of all types serve as some of the most effective user interfaces, especially when computerized.

Digital mapping technology has been in use since the early 1980s, however, there has been a lack of application to marketing analysis.[67] ERSI was an early leader in the field, concentrating in the urban planning and environmental science areas.[68] In most situations, applications of digital maps have involved proprietary systems and are stand-alone projects. Likewise, the ability to combine data sets together for input into mapping software is also dominated by proprietary systems that take the form of software packages.

In these situations, there is no capability to link data to mathematical models for improved analysis. This lack of an open system, Internet based approach has slowed the progress of mapping and graphics applications use in day-to-day decision-making. With advances in Web Services (XML and SOAP) along with the advent of browser based open systems this situation is changing.

In addition, advances in RFID technology provide another means of obtaining consumer data with a geographical component. The read event for an EPC, which contains "what," "when," and "where," represents a parallel stream of data, in addition to loyalty and credit/debit cards (see Figure 15-3). As the costs of the technology decrease and reliability improves, it is probable that this form of data collection will surpass financial transactions and loyalty cards as a means of gaining geographic specific data on new products. In any case, the use of standard protocols to move data across application interfaces will enable improved analysis. Combining this approach with visualization is essential in helping managers make quick decisions.

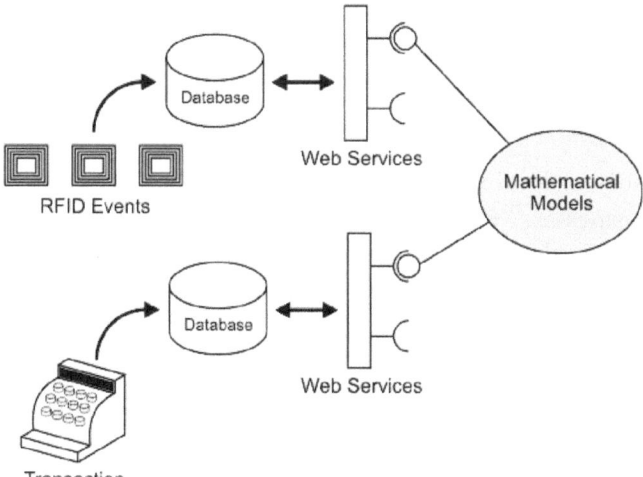

Figure 15-3 – Gathering Data for Spatial Diffusion Analysis

Diagram credit – Ching-Huei Tsou and Professor John R. Williams – MIT Auto-ID Labs, Used with permission.

Important Developments in Digital Mapping Technology

The City of Philadelphia announced in February 2005 an interesting project to build a 3D model of the downtown area.[69] Using aerial photographs and ground imaging, the representation provides stunning detail of buildings and shops even to the level of mailboxes and shrubbery. This physical representation can be combined with other data such as office rental vacancies to provide a virtual world accessible from any computer. In a similar effort, Yahoo is offering a new service that provides maps of traffic conditions for more than 70 metropolitan areas based on real-time data from embedded road sensors, traffic cameras, police scanners and traffic helicopters.[70]

Google has also entered the digital mapping market with the purchase of Keyhole in October 2004. The Google technology uses aerial photographs at different resolutions along with an existing browser technology collectively termed Ajax (Asynchronous Javascript and XML). This allows users to zoom in through aerial maps at different resolutions to street level images that show details of buildings and urban areas. The Ajax technology uses Javascript, dynamic HTML and XMLHTTP.[71] This set of technologies provides speed and responsiveness in viewing digital maps. The major advantage of using Ajax is that it allows asynchronous data transmission between clients and server, so users can interact with the program without waiting for the "postback," which occurs every time a button is clicked on a traditional web page. All of this technology is open source, allowing programmers to write applications quickly.

Google even provides for local map search capabilities, allowing users to locate individual businesses.[72] Plans include linking local search to potential advertisements about the business, essentially becoming an alternative to locally published Yellow Pages.[73] This is an important development because according to a national survey of 99 firms conducted by Forrester Research, 85% plan to increase online advertising spending.[74] The market for online advertising continued to expand at a double-digit rate through 2004 and 2005.

Taking digital mapping a step further, several emerging businesses extend the technology developed by Google and others to plot real estate for sale on maps along with overlays of schools, access to main roads, restaurants, medical offices, and coffee shops within the vicinity. Through a single glance, potential buyers can determine if a particular home or apartment meets their needs in terms of convenience and life style. An example can be found at www.housingmaps.com.[75]

With these emerging technologies, there are opportunities to integrate digital maps, spatial data, and mathematical models into entirely new tools for analysis. In the next section, several examples are put forth to highlight the potential of using digital mapping technology as a tool for analyzing spatial diffusion.

Using Aerial Images in World Wind

The most common means of plotting data for spatial diffusion studies involves paper maps. Figure 15-4 is an example of a paper map used to show spatial diffusion differentiated by innovators, adopters, and non-adopters.

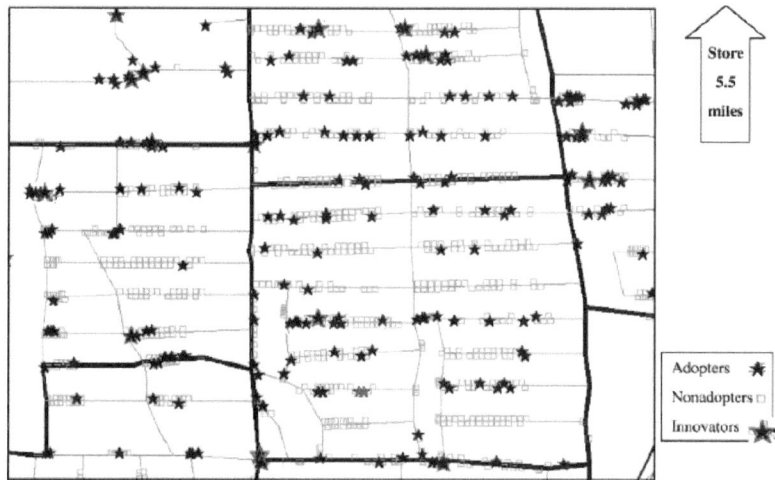

Figure 15-4 – Mapping Spatial Diffusion

(Allaway, Berkowitz, and Giles 2003)
Used with permission, Journal of Retailing

Though this means of visualization is effective, various types of three-dimensional digital maps carry more information about the geographical area under study. For example, Figure 15-5 shows an aerial map of the MIT campus viewed through World Wind, an open source application developed by NASA and written in C++ (Microsoft .Net) and Managed Direct X (http://worldwind.arc.nasa.gov/). The application uses aerial photographs from the US Geological Survey along with topographic maps from USGS and satellite images taken from NASA Landsat 7, a satellite system. At some time in the future, real-time satellite images might be available.

Figure 15-5 – Aerial Picture of MIT

Diagram credit – Ching-Huei Tsou and Professor John R. Williams – MIT Auto-ID Labs,
Used with permission.

Because this imaging technology is open source, World Wind can be adjusted to overlay data using Web Services interfaces to various data sources. As long as a data source is exposed through a Web Service, the application (World Wind) can consume the data. This adds additional value to the map. Some public sources for additional data overlays might include:

1. Geocode from the US Census Bureau termed Topologically Integrated Geographic Encoding and Referencing (TIGER).

    http://www.census.gov/geo/www/tiger/

    The information about spatial objects is organized in records files. The 2004 TIGER/Line data consists of 19 record types that collectively contain attributes like address ranges and ZIP codes, street names, and latitude and longitude coordinates.

2. Conversion between mailing addresses and longitude, latitude.

   TIGER/Line data is publicly available. It is in text file format. By storing the data within a database and exposing through web service, TIGER allows for mapping between mailing addresses and longitude/latitude.

   http://pdait1.mit.edu/Geocoding/lookup.asmx

3. Zip Code Tabulation Areas
   This provides the boundary (lat, long coordinates) for each Zip code area.

   http://www.census.gov/geo/ZCTA/zcta.html

   Utilizing these resources, such things as zip code boundaries can be added to digital maps (Figure 15-6). The approach can also be used for aerial maps (Figure 15-7). With the aerial maps, the user has the ability to zoom into a particular area and see details such as the type of buildings and transportation network for an urban area.

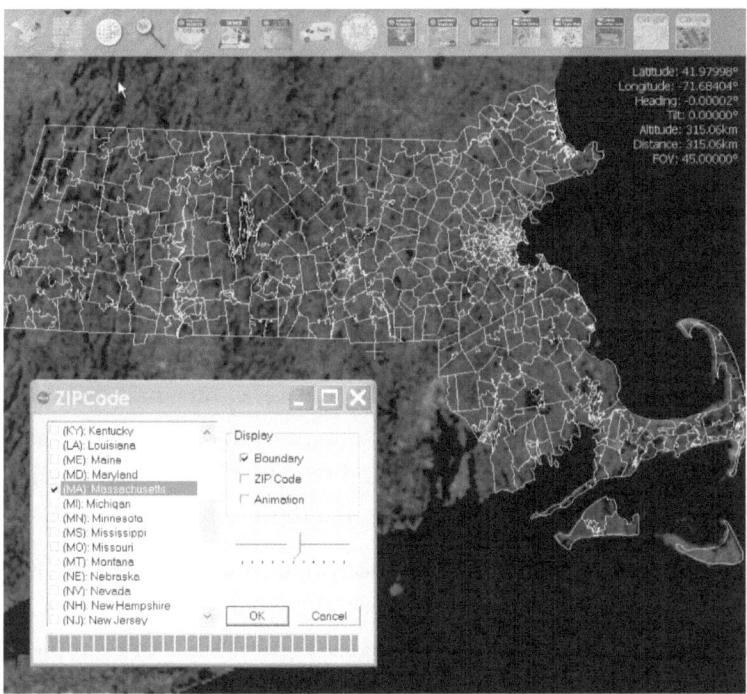

Figure 15-6 – Zip Code Boundaries for Massachusetts

Diagram credit – Ching-Huei Tsou and Professor John R. Williams – MIT Auto-ID Labs, Used with permission.

Figure 15-7 – Zip Codes Overlaid on Aerial Map

Diagram credit – Ching-Huei Tsou and Professor John R. Williams – MIT Auto-ID Labs, Used with permission.

Interoperability of Data

Taking advantage of the open sourced architecture of World Wind software, other digital maps from providers like Google can be used. It is also possible to use USGS maps in the Google Maps Web application. This type of interoperability between different data sources results in a great deal of computational flexibility. It should be noted that this sometimes requires web applications to convert different scales and datum measurements.

The Google mapping software includes both two-dimensional maps and satellite images from Keyhole. In some cases, the Google maps provide more detail as compared to the USGS images (Figure15-5). An example of each is shown in Figures 15-8 and 15-9 (both show areas of MIT along the Charles River).

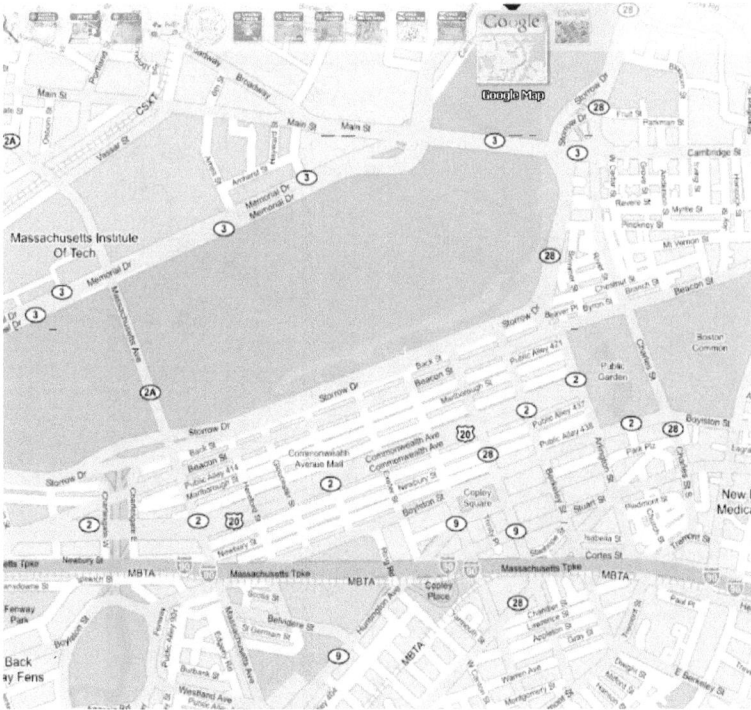

Figure 15-8 – Google Two Dimensional Maps

Diagram credit – Ching-Huei Tsou and Professor John R. Williams – MIT Auto-ID Labs, Used with permission.

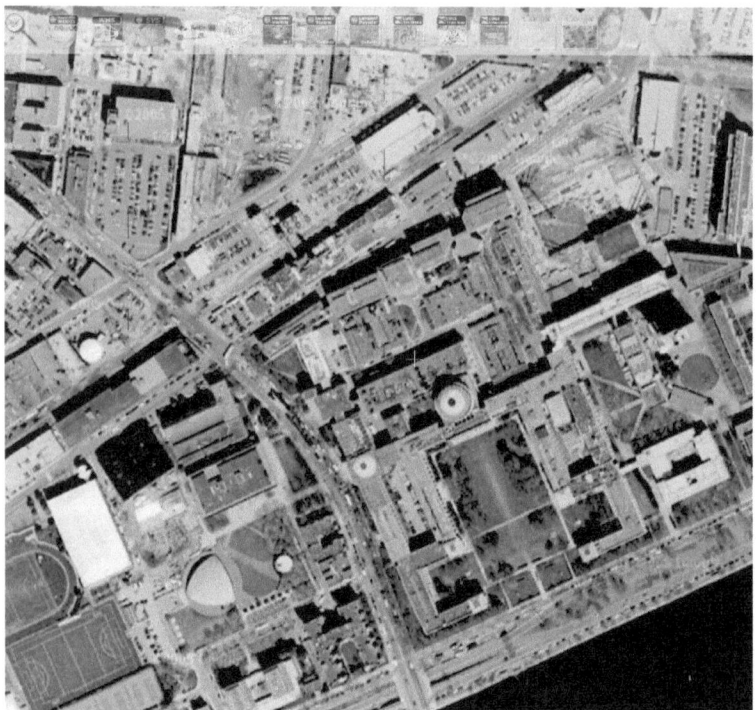

Figure 15-9 – Google Aerial Map of MIT

Diagram credit – Ching-Huei Tsou and Professor John R. Williams – MIT Auto-ID Labs, Used with permission.

Spatial Diffusion

Another example of this digital mapping technology involves the visualization of spatial diffusion for a particular product. The interesting aspect of this example is that diffusion can be animated to show changes through time using different types of maps and at different resolutions. If street address information is available, adopters can be plotted automatically as part of the animation. As an alternative, if only zip code data exists animation of spatial diffusion can still take place, although at a lower level of resolution. Even animations with low-level resolution through time have value if correlated with aggregate demographics for a particular area (Figure 15-10 and 15-11).

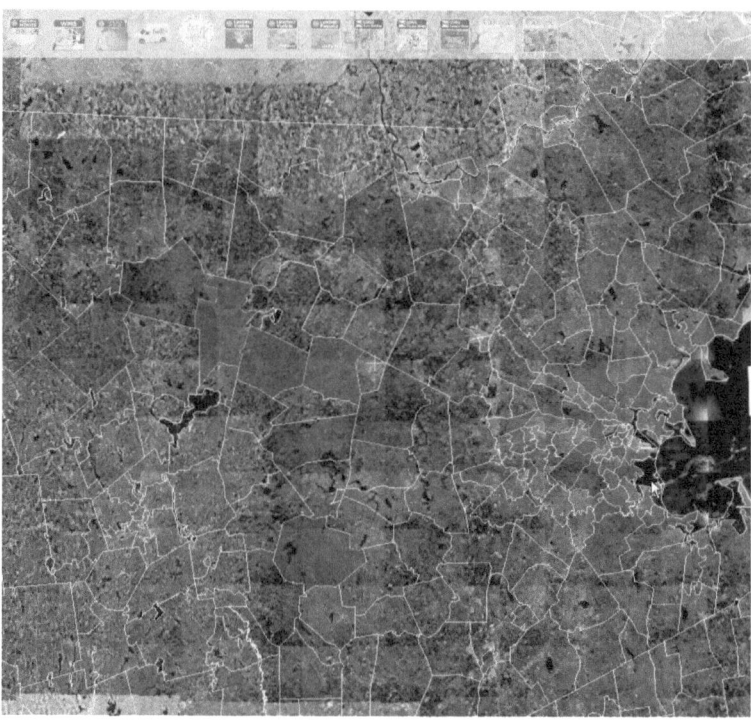

Figure 15-10 – Spatial Diffusion, Aerial Map

This diagram represents a digital view of spatial diffusion. The view can be simulated through time.

Diagram credit – Ching-Huei Tsou and Professor John R. Williams – MIT Auto-ID Labs, Used with permission.

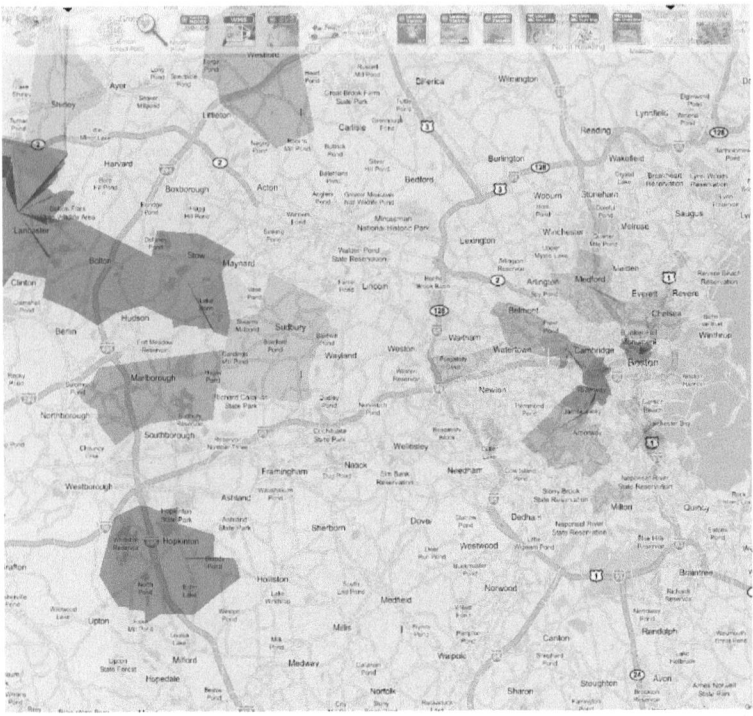

Figure 15-11 – Spatial Diffusion, Two Dimensional Map

This diagram represents a digital view of spatial diffusion. The view can be simulated through time.

Diagram credit – Ching-Huei Tsou and Professor John R. Williams – MIT Auto-ID Labs, Used with permission.

With improved digital mapping technology, open source systems,[76] data, and model interoperability, detailed studies of spatial diffusion are possible at the street address level for new product launches. Assuming compromise can be reached on privacy concerns, spatial diffusion will become an important tool in new product demand forecasting and optimizing advertising expenditures during an initial launch.

Making this tool a part of everyday management decision in supply chain and marketing will require continuing integration of data, digital maps, and mathematical models. The information technology infrastructure is in place to accomplish this integration with the long-term prospect of improved new product launches in terms of supply chain efficiency, cost, and success.

Knowing the location of customers is perhaps the most important part of retailing.[77] This is particularly true in choosing the means of advertising. The announcement of Google's proposal to offer city wide, free WiFi access for San Francisco highlights the importance that the company places on customer location. As part of the announcement, a company representative noted that Google, "…is especially interested in testing out future services and advertising systems that take advantage of knowing a user's location, which would be possible through offering such wireless Internet access." [78]

The emphasis regarding customer location will only increase in the future as innovative companies like Google develop new approaches in advertising. Hence, the EPC will play a greater role in customer location, interactive marketing, and spatial diffusion analysis.

CHAPTER 16

Outlook: Navigating the Sea of Data

At Smart World 2004,[b] Sunil Gupta of SAP paraphrased Samuel Taylor Coleridge by saying "data, data everywhere but not a byte to use." [1]

Each year, the amount of data grows by as much as 40 – 60% for many organizations.[2] In 2004 alone, shipments of data storage devices equaled four times the space needed to store every word ever spoken during the entire course of human history.[3]

The explosion of storage device sales has gone largely unnoticed because the selling price has decreased sharply during the past few years. While the increase in unit sales has meant larger revenues for companies like EMC, the amount of the increase has not been great enough to draw widespread attention. Since the growth rate appears to be exponential, businesses are likely to face a mounting data organization problem in the years to come.

Compounding matters, a large amount of the increase is from unstructured data that are extremely hard to organize using current computer-based approaches. Unstructured data includes images, text, emails, and engineering designs.[4] In all of these cases, object representation requires more than a serial number that can be stored in a database.

For example, a serial number does not adequately describe data such as an image or a section of text. It is only through using words in a machine understandable way that descriptions of these objects can become useful for search, organizational, and analytical purposes.

Amidst the ever-increasing amounts of data, "…companies are struggling to figure out how to turn all those bits and bytes from a liability into a competitive advantage."[5] Emerging technologies such as the EPCglobal Net-

---

[b] Smart World 2004 was sponsored by the MIT Industrial Liaison Program

work and RFID technology, along with sensor networks and the use of "loyalty cards" are certain to generate even more data.

By one estimate, there will be 300 million RFID readers active in supply chains within ten years.[6] In another prediction, "the number of deployed sensors will dwarf the number of personal computers by a thousand fold" in 2010.[7] Both of these technologies will boost the amount of raw data available for supply chain management.

Dealing with the increasing volumes of structured and unstructured data will require new standards and information architectures to improve integration and communication between hardware, software, and business entities. This becomes important as companies seek to overcome the barriers that limit the seamless transfer of data, internal and external to the firm. However, the bigger question remains "How are we going to make sense of large volumes of data generated from RFID?"

Analyzing Data Using Models

The underlying success of supply chain management depends on the flow of data and information for improved business decision-making.[8] Unique identification, interoperability, standards, and automated Internet-based systems to track, trace, and control physical objects are all important elements of RFID technology that are moving out of the laboratory and into practical application within various commercial and governmental supply chains. Though there is a long road to full implementation of the EPCglobal Network, the merging of information and physical objects opens so many new opportunities that it is important for all organizations to plan for future operations by learning as much as possible about the technology.

Analyzing the large volume of raw data (including real-time telemetry) produced by RFID and sensors in an orderly way requires the additional use of mathematical models to provide understanding. Mathematical models are simple representations involving characteristics of the real world that are determined to be important.[9] Models highlight facts and interests at hand, and depict only part of reality. Some go as far as to say that the human thought process is essentially a specialized model of the real world.[10]

Mathematical models are especially useful in making sense of complex situations like those encountered in supply chain management. Beyond identifying important issues and serving as an aid to communication, models provide the greatest value in suggesting explanations for observed events such as unanticipated build-ups of inventory or poor customer service.

Though mathematical models are extremely useful in providing insight, the process of building models often lacks productivity because development seldom follows a linear path, [11, 12] and because separate natural, mathematical, and computer representations are needed for managers, model builders, and computer programmers.[13] This increases the need for detailed interfacing that tends to inhibit seamless sharing of models within a network. As a result, implementing mathematical models is complex, time consuming, and requires advanced technical capabilities and infrastructure.

Although there is a strong history of applying models to help managers make decisions about intricate supply chains, specialists often develop these comprehensive models internally within business organizations or academia. This is commonly an application specific job and the same model building technique must be re-invented afresh for each new situation. Though internal development can lead to significant breakthroughs, this approach depends on trial and error to find what works in practice, combined with mathematical intuition and an extensive knowledge of technical publications.

In the subsequent sections of this chapter, the initial standards for a new computer language are put forth that will enable computers to describe and share models across the Internet and to interoperate data obtained through RFID technology [14, 15] and other means. This will substantially increase the Clockspeed [16] of modeling, and the computational efficiency of applying models to perform the functions of "sense," "understand," and "do" that comprise the underpinning of creating smart objects within supply chains.

Building a Network Using Words

At the most basic level of communication, words are the glue that connects nearly everything together. The power of words can give descriptive meaning to the most complex physical objects existing in business or nature, and to the most diffuse ideas that exist only in the mind. Data is described with words, information is word based, and computer code uses words as a way of communicating the various operators available to programmers. Words establish not only the limits of human imagination and intellect, but also the possibilities for the computing systems of the future.

Every word has at least one definition and, when used in conjunction with other words, it is possible to create a countless number of sentences. For example, a simple cartoon shown to twenty-five different people will generate twenty-five individually unique perspectives if each writes a single sentence about what they see.[17] Dictionaries organize and define words used to make sentences, and a weak ontology[c] based on groupings such as synonyms and antonyms provides the relationships that the human mind can fathom into linguistic communication.

In spite of the almost limitless capacity of human language, describing physical or abstract objects in a consistent machine understandable way remains a challenging objective for both computer scientists and practitioners within business. Although English is a powerful tool for communicating meaning through words, noun phrases, and sentences of varying patterns and complexity, this ability is under-utilized by modern computing systems because meaning is a combination of syntax, semantics, and context that is beyond the cognitive abilities of computers.

The fundamental problem with employing words as a descriptor is that a single word can have several different definitions and multiple words can have the same definition. This paradox means that natural language often does not have the internal consistency required for straightforward application as an identifier or a unit of meaning within computer systems.

Complicating matters, the intricacy of meaning increases dramatically when dealing with the noun phrases and sentences needed to describe abstractions. Given this property of English, it is impossible with current technology to conduct a semantically precise, computer-based search of information contained in web pages (HTML), quantitative data tagged with words, news feeds comprised of text files, complex mathematical models, or any other situation where words describe physical or abstract objects. Achieving the goal of word descriptions that are machine under-

---

[c] In computer science, an ontology represents the relationships between things.

standable requires a deeper appreciation of the role of semantics in computing systems, and especially the Internet.

Integrating Semantics into Computer Communication

The Internet provides a natural means of sharing data and models between organizations. However, the existing standards of the Internet do not provide any semantics to describe models and data precisely or to achieve interoperation.

For the most part, the Internet is a "static repository of unstructured data" that is accessible only though extensive use of search engines.[18] Though these means of finding data have improved since the inception of the Internet, human interaction is still required and there are substantial problems concerning semantics. In general, "HTML does not provide a means for presenting rich syntax and semantics of data." [19]

For example, one of the authors of this book did a search for "harvest table, oak" hoping to find suppliers of home furniture. Instead, the search yielded a number of references to forestry and the best time to harvest oak trees. Locating the URLs relating to furniture required an extensive review of a number of different web sites. This process of filtering can only be accomplished though human interaction and is time consuming.

With inaccurate means of doing specific searches based on one semantic interpretation of data, information, or models, it is nearly impossible for the Internet to advance as a productive tool for modeling the vast quantities of data resulting from RFID technology.

Several Types of Webs

The problem of semantics arises from the fact that keywords are the means used to describe the content of web pages and other material that appears on the Web. As noted, each keyword can have multiple meanings, creating a situation of great difficulty when attempting to accomplish an exact search. The difficulty increases by an order of magnitude when attempting to do phrase-based searches. Without exact search capability, it is impossible to create any sort of machine understandable language for the current *Web of Information.*

Even though the search engine issue has not been resolved, the EPC and RFID technology lay the groundwork for a new type of Internet characterized as the *Web of Things*. With serial identification of physical objects utilizing the EPC, searches accomplished through Internet search engines or proprietary IT infrastructures will become much more effective in finding an exact match.

This provides the ability to do track and trace across entire supply chains and other computerized functions important to business. A strong case exists that linking the physical world using RFID technology and ubiquitous computing will form the basis for a revolution in commerce by providing real-time information and enabling smart objects.[20]

As impressive as the effort to create the *Web of Things* has become, it still does not address the question of semantics in describing objects beyond the use of a simple serial number. A large number of abstractions, such as mathematical models and data, cannot be characterized by a unique serial number no matter how sophisticated the syntax. Without the ability to provide unique identification of an abstraction, the Internet will serve little useful purpose in linking mathematical models and data together in a way similar to the manner that the *Web of Things* will eventually link the physical world.

In the future, the means of semantic description for models will become extremely important for the analysis of RFID generated data. To accomplish this higher goal, the Internet must become a *Web of Abstractions*, in addition to a *Web of Information* and a *Web of Things*.

Through a *Web of Abstractions*, models can be matched much more quickly to practical problems, along with the available data, and shared beyond single end-user applications. This capability is of great value to practitioners who are interested in gaining the maximum understanding of data generated from RFID technology.

The general mechanism of creating a web of abstractions is Semantic Modeling, an overall approach that allows for precise descriptions expressed in words that are machine understandable. Through Semantic Modeling, practitioners will be able to match models to data in a timely manner with greater productivity and relevance.

The vision of Semantic Modeling calls for a network where data from widely divergent sources merge seamlessly into a coherent whole within and across enterprises. In this new type of network, mathematical models, algorithms, and software can automatically combine with data to form an intelligent information infrastructure.

From a practical viewpoint, there are ten basic areas where Semantic Modeling provides value to business. These include:

– Creating a standard format for data that allows sharing
– Internet search using the definition of a word
– Various forms of data visualization
– Improved data quality
– Basic forms of automatic language translation
– Linking models and data within a network
– Aggregation of data
– Standards for spatial data
– Standards for data from sensors
– Standards for location

The specific tool to accomplish Semantic Modeling in practice is the M Language,[d] a comprehensive means of achieving large-scale data and model interoperability. Though this language is still under development, the general direction of improved interoperability remains clear. For organizations to realize the full value of RFID, new technologies must emerge that will eliminate barriers to the flow of data along with the ability to deal with semantics in a machine understandable way. The M Language accomplishes this task.

The next two sections of this final chapter outline the details of a new approach to standards along with a discussion of the M Language.

Establishing a Standard for Semantics

The current solution to the problem of semantics involves relying upon various standards groups that spend a great deal of time attempting to derive a single, universal definition for a commonly used word or noun

---

[d] The M Language is currently under development at the MIT Laboratory for Manufacturing and Productivity – Data Center Program, www.mitdatacenter.org.

phrase that fits all contexts within a particular industry. Several groups have gone as far as to employ the practice of creating camel case words, a situation where two or more words combine to form a new word. The new camel case word then takes on a single meaning. This practice expands the number of words used to describe objects; however, various industries might still assign different definitions to a single camel case word. In addition, a loss of flexibility occurs when combining strings of words into new words.

Some examples of standards groups include RosettaNet™ (a division of GS1), the National Retail Federation (NRF), Association for Retail Technology Standards (ARTS), the GS1-US and GS1 (GS1 US formerly known as the UCC), Global Data Dictionary (GDD™, a product of GS1) and many others. Though not participating in formal standards development, nearly all non-profit professional groups establish a dictionary to give common meaning for frequently used words. Several consortia like RosettaNet go a step further in developing standard terminologies and in some cases use XML-based schema as a means of describing business processes and data, and other abstract objects commonly used in commerce.

Though the single definition method has worked very well in highly structured situations, each industry segment tends to become unique regarding the words and definitions employed. This insular approach increases transactional efficiency within a particular industry. However, industry specific definitions also sacrifice opportunities to share data, models, and other abstractions across industries because of the lack of a universally agreed upon definition that is globally visible.

Limits of the Standards Approach

Though developing a single definition for words or noun phrases works well within science and engineering, there are significant limitations when applying this approach to business situations. Researchers have employed various computer-based techniques, such as Artificial Intelligence (AI), in an effort to infer the contextual meaning of words associated with a physical or abstract object.

These approaches have largely failed in practice because the meaning of a word, a phrase, or a sentence used to describe an object depends on the semantics of each word, the syntax or order of usage, and the context in which the word(s) appear. The unique properties of the human mind can determine contextual relevance, and then figure out how several relevant variables are associated.[21]

To date, machine languages have not duplicated this human property. Further, most AI techniques rely upon deductive reasoning where general concepts are applied to solve specific situations in terms of meaning. This becomes difficult to do in practice, because most word meanings in business situations are inductive with a strong dependence on specific context.

Interoperability in Practice

The current inability to interoperate and search data, information, and models arises from the need for translation between various independent dictionaries (situated in numerous standards groups), which involves writing a custom computer program for each translation. This situation becomes even more complex as the numerous dictionaries currently in existence undergo revision as words change meaning over time.

For example, the word "Chad" has existed for 14 centuries as the name of a saint who lived in the most northern county of England.[22] During the U.S. Presidential election of 2000, the word took on new meaning where a hanging chad referred to an improperly punched ballot.

As commerce expands, the use of distributed dictionaries will become even more unmanageable. By one estimate, for every n words contained in existing dictionaries there need to be about n^2 translations as part of normal communication within industry.[23] As the number of words used to describe objects in a machine-understandable way increases, and dictionaries become larger, the volume of translations will become untenable, thus making inter-industry data sharing impossible to achieve in practice.

In the area of computer transaction systems, single-term standardization has its roots in several hundred years of engineering standards development, where the historic goal was precision in definition for highly specific situations along with universal adoption of the agreed upon definition. In the case of commerce, most of these standards efforts have focused on words used in transactions with little emphasis placed on describing complex objects such as data, information, business processes, and mathematical models in a common, interoperable way.

Establishing Semantic Based Computing

The essence of developing future machine understandable semantics depends on the fact that language, and English specifically, is relative in meaning based on the intended usage and context by those who originate the communication, object description, or search. This becomes apparent when comparing the definition of words used in different business and academic disciplines.

Since all words have definitions residing in various dictionaries, and all words are subject to classification, it is possible to design a computer system that utilizes multiple definitions for a single word, yet maintains system-wide consistency in relative meaning. Though complete machine contextual understanding of sentences is a distant goal, current computer technology, given the proper architecture, is capable of applying relative meaning to words and noun phrases used as descriptors for abstract objects like data, information, and mathematical models.

Relying on words and noun phrases as descriptors with multiple meanings defined in a centralized open dictionary, drastically improves the prospects for finding an exact semantic match when conducting an Internet-wide search based on a single definition. This offers the possibility of searching and matching data, information, and models from different disciplines, creating new types of interaction that do not currently exist because of barriers in description, definition, or format.

The last section of this chapter provides details of the M Language, a means of incorporating semantics into computer-to-computer communication.

A Language for Interoperability

The M Language is conceptually simple, consisting of two parts, *words*, and *rules*. In M, words take on a new form that allows for machine understanding. Rules provide guidelines about how to place words together for representing data or models in a common format that is interoperable. These representations are in the form of messages capable of transfer between computing systems within an organization or across the Internet.

Words and rules are the starting point for understanding the M Language.

Words

The words used in the M Language are slightly different from English words. In M, every word has only one definition. This is an extremely important characteristic because computers that communicate using M do not need to understand the context or usage of a word to know its meaning.

English words are ambiguous. For example, the word "cell" might mean "cellular phone," "biological cell," "jail cell," or "fuel cell." Without some idea of the context, it is impossible to know the meaning of the word "cell."

To overcome this issue, the M language includes a number to denote individual words as in the following example:

    cell.1

To account for multiple definitions, the M Language allows numeric extensions, one for each definition. Thus, cell.1 is a word in M and cell.2 is a different word. With this method, every word has only one meaning.

In English, dictionaries define the meaning of a word. M also uses a dictionary. The M dictionary serves as a repository for definitions of words used in computer transactions. The dictionary also is a means of storing other important information associated with a particular word. This provides an effective means of unifying various aspects of a word and forms a base for common computer-to-computer communication.

The M Dictionary

In the M-Dictionary, words and definitions are stored in the following form:

cell.1	The basic structural and functional unit of all organisms; they may exist as independent units of life (as in monads) or may form colonies or tissues (as in higher plants and animals).

This, of course, is the only definition for the word cell.1. Other words, such as cell.2, cell.3, and cell.4 all have different definitions expressed using the same format. Depending on usage, some compound words are also part of the M Dictionary. An example is "operations research," represented as operations_research.1.[24]

In addition to the definition, the dictionary entry also contains three other pieces of important information. These include (1) word relations, (2) data format, and (3) language translations. This information helps in forming and understanding messages composed in M.

Word relations are simply the connections between words. These relationships include synonyms, antonyms, types, and parts.[25] Synonyms and antonyms are the same as in English.

Types refer to word generalizations. For example, automobile.1 is a *type of* motor_vehicle.1.

Parts are words that are components of another word. This is often the case when thinking about physical objects, although this could also be the case with abstractions. For example, a wing.4 is a *part of* airplane.1. [26]

Both types and parts establish a hierarchy within the dictionary through making connections between entries. These word connections are valuable in a number of different ways, including improved search.

Data format provides guidance concerning the forms and patterns of data values that are associated with a particular word. In many situations, computer-to-computer communication might contain a word such as first_name.1 that has an associated data value such as "John." Other common situations include words like telephone_number.1, account_balance.1, or postal_code.1. In all of these cases, a word in the dictionary has a particular format or pattern for associated data.

Finally, the language translation portion is simply the representation of the word in M as a word (or phrase) in another human language besides English. In most situations, computer-based language translation is very difficult because a lack of context exists for the specific communication. Since in M each word has only one definition, the word cell.1 (biological), for example, cannot be confused with cell.2 (telephone). Words with a single definition allow users to specify exact meaning independent of context. This eliminates ambiguity in translation.

Dictionary Development

Developing common definitions for the words, data formats, and translations used in commerce along with the analysis of data across all industries has traditionally been a source of great debate within business. To build a robust global dictionary containing the words used in the M Language along with other important information, such as relations between words, associated data patterns, and translations requires a different approach.

The "wiki" process has emerged as an innovative application of Internet technology to knowledge management and consensus building. A 'wiki' is a type of website that allows users to add and edit content and is especially suited to collaborative authoring.[27] It is remarkably accurate [28] and several companies have begun to use the process internally.[29]

Since 2001, Wikipedia has become the largest encyclopedia ever created with over 3 million entries.[30] The M dictionary uses the wiki approach with several important modifications including improved security through user registration, maintenance of the integrity of word relations, a monitoring function to reduce the chances of near identical definitions, and administrative controls to ensure accuracy. The dictionary also has various statistical features that measure usage.

However, having a robust dictionary is just a part of the M Language. To form messages, computers need a set of rules that give instructions on how to glue the words together. The next section discusses the rules of the M Language.

Rules

Language is more than just a collection of words defined by a dictionary. For most languages, grammar gives explicit rules concerning the word order needed to give meaning to a sentence. In English, the simple sentence "Threw ball the Jack" has little meaning. Establishing correct word order is essential. Thus the sentence "Jack threw the ball" formed by rearranging the words makes sense.

From this example, it is clear that word order, sometimes called syntax, has an important role in communicating meaning. If words are in the correct order, instant recognition takes place.

Just as English has rules of grammar for word order, the M Language also has rules establishing the order needed for machine understanding of messages.

The initial version of the M Language contains three simple rules. These three rules, however, represent a significant portion of computer-to-computer communication. The three are (1) phrases, (2) key-value pairs, and (3) tables.

A phrase is a sequence of machine-understandable words representing a single idea. A phrase in M is just like a phrase in English. The syntax is such that the last word in the phrase is the root and all the others are modifiers. As an example, the phrase "initial account balance" appears in M as:

```
initial.1_account.1_balance.1
```

In this phrase, balance.1 is the root word, while initial.1 and account.1 are modifiers. Phrases within the M Language represent a unit of meaning that is extremely useful in increasing the precision of data element descriptions.

Key-value pairs are simply a list of words with associated data values. Tax forms, medical records, and financial statements are all representations of key-value pairs. An example is:

```
Name.1 — "John Smith"
Telephone_number.1 — "(211)-459-1234"
```

Key value pairs in the M Language are useful in making data interoperable within and external to the firm. Interoperable data opens a number of possibilities for combining data posted on the Internet with internal company data.

The final rule involves tables, which are the most common way to store data on the computer. There are many different ways to represent tables; comma separated values (CSV), Excel spreadsheets, HTML tables, and others. In the M Language, a table takes on the pattern of repeating sets of key value pairs, each with identical keys. The following is an example:

```
patient.1
    name.1 — "John Smith"
    telephone_number.1 — "(211) — 459 — 1234
patient.1
    name.1 — "John Doe"
    telephone_number.1 — "(211) — 459 — 4321
```

Subsequent versions of M will include additional rules for spatial data, equations, and mathematical models. Integrating spatial information, for example, will be valuable in marketing, demographics, transportation, and logistics, while rules for mathematical equations and algorithms will ultimately allow the integration of data and models.

Applications

The M Language is a tool that enables the free flow of data and models across the network by creating machine understandable semantics and syntax. Achievement of this vision will result in a number of practical applications in industry. The final section of this chapter outlines a few prototype applications.

Interoperable Data

Perhaps the most obvious application of the M Language is as an intermediary between proprietary data systems. In this application, data from one database is translated into M *before* it is communicated to another, as shown in Figure 16-1. Here data from a source system is translated *at the server* into M. This data is sent via the network – either as human readable text or compact binary – to the target system. The data can then be used in the native M format, or translated into another proprietary schema and stored in the local database.

In broad terms, the M Language serves as a common transport between distributed, incompatible data systems. The advantage of this approach is that data providers do not need to know the format or content of every possible target application. Providers need only expose data in the standard M language for data to be interoperable. Using M as a common carrier, translation takes place only once at the server instead of many times for every possible consumer.

This approach is a powerful means of merging data available on the Internet with private data resident within a firm. The resulting data set would be under the control of the firm and not be open to the Internet.

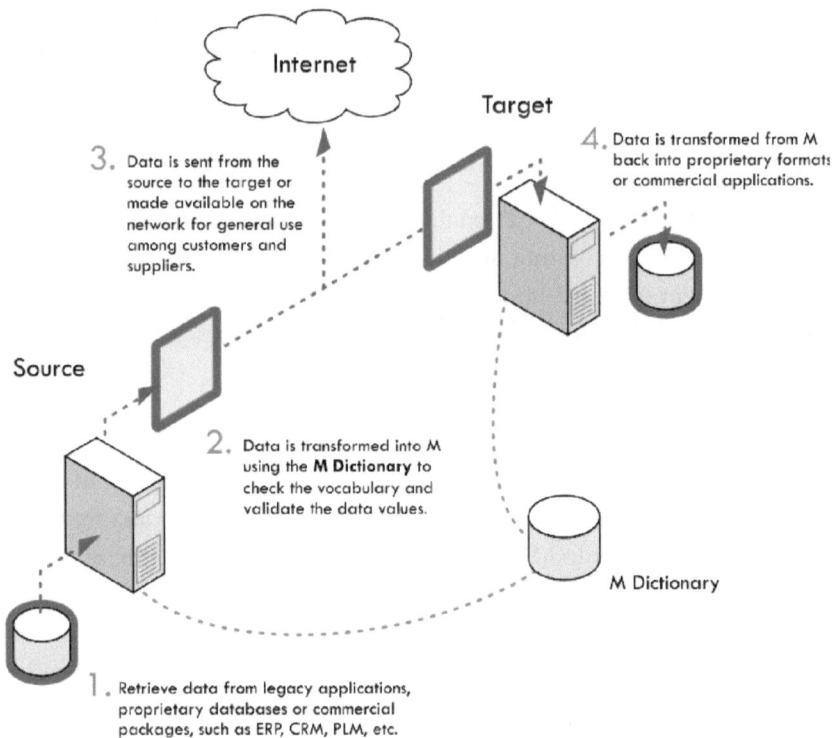

Figure 16-1 – The M Interface

Browser

The M Language provides more than just a data translation service. As a common language, M encourages third-party developers to create a wide range of software tools and applications. As a first step, the MIT Laboratory for Manufacturing and Productivity - Data Center Program has created a browser to view, edit, and manipulate data directly in M. The browser presents data in an easy-to-understand format without the need for additional styling information. Sections are indented, headings aligned, data color-keyed and tables displayed properly as spreadsheets.

The application also provides data plotting functions, definition based search, data validation, language translation, and model integration. Data plotting functions refers simply to charts and graphs common in spreadsheet applications. The other features, however, are novel.

Definition Based Search

Since words in M have only one meaning, a key-word search in M yields matches based on the definition, not on the character string. This allows for search based on meaning rather than keyword. Further, the dictionary proposed for M provides additional search capabilities as compared with current approaches.

Every definition in the M dictionary includes provision for word relations. From the previous example in this chapter, automobile.1 is a *type-of* motor_vehicle.1 and wing.4 is a *part-of* an airplane.1.

Combined with definition-based search, these relations provide a powerful tool. For example, search can span types of a word. A query for types-of flower.2 would return rose.4, violet.3, or marigold.1. A search for parts-of automobile.1 would return fender.1, muffler.3, or engine.1.

Searching on a definition and using the word relations from the dictionary greatly increases the precision of Internet search. In addition, words and phrases from the M dictionary could become part of meta-tags used in HTML, significantly increasing the accuracy and performance of web-based search tools such as Google, MSN Search, and Yahoo.

Data Validation and Quality

Given the rapidly increasing amounts of data available within business, data integrity is taking on even greater importance. In addition to definitions for a word, the M dictionary also specifies the structure and format of any attached data value.

For example, the entry telephone_number.1 might specify that data values should conform to those of a specific country or internationally recognized patterns. Thus, an entry (617) 258-123 would be recognized as incorrect since United States domestic telephone numbers require a four digit extension. Likewise, the entry 128(0) 2522 1111 is also improper because a country code of 128 does not exist.

This capability provides an opportunity for data sharing and information validation at time of entry, with the goal of improving data quality.

Language Translation

Another benefit of the M Language is translation to different languages. Since words in M have only one meaning, translation to another human language is much easier. It is not necessary to use the context or meaning of a passage to discern the definition of a word.

For example, cell.2 means cellular telephone. Thus cell.2 translates into "cell phone" in American English, "mobile" in British English, "Handy" in German, "portable" in French, "手機" in Chinese, or "джиесем" in Bulgarian.

Model Integration

The most far-reaching feature of the M Language is the ability to share not only data across the network, but also models. All mathematical models contain three basic features – inputs, outputs, and an algorithm for computation.

By describing the inputs and outputs in the M Language, a model can be linked to data semantically. In other words, because the data and models are described in the same language, they can be connected together using the words and word relations from the common dictionary.

Models can also be connected to one another. In this case, outputs from one model can be matched semantically to the inputs of another model, forming a model pair. Connecting more models in the same way can build the simple pair into a larger *model network*. These model networks can then function as single units, operating on data and performing complex mathematical analysis.

The Future

As Professor Grosof of MIT once commented during a public lecture, there are only several certainties in the world, "death, taxes, and integration." [31] Interoperability through greater semantic integration will be one of the major IT trends of the next ten years.

Even before the advent of computer based communication and networks, scholars from many different disciplines have often noted the importance of language and the value of a universal dictionary to the overall health of nations and to the advantage of commerce. Samuel Johnson, the creator of the first English dictionary, once commented, "Languages are the pedigrees of nations." [32]

In the future, languages for computer-to-computer communication will become the pedigrees of businesses. To this end, the MIT Laboratory for Manufacturing and Productivity - Data Center Program is actively pursuing the research, development, and application activities needed to build the next generation of computer systems that will connect physical and abstract objects through words and noun phrases of different definitions. Much of this work involves integrating existing standards developed by the W3C[e], such as XML, with fresh innovations. This new generation of semantic-based systems will be capable of organizing and searching the ever-increasing volumes of data within business.

Navigating the sea of data will be one of the most important aspects of gaining success from the EPCglobal Network and RFID technology. This requires skill along with the innovative spirit that flourishes in a free market enterprise system. Meeting the challenge of implementation is a tremendous task with an uncertain outcome. The only certainty is that the future will hold failures along with successes, and unforeseen events will cause changes in course. Thriving in this environment requires exceptional ability and great knowledge of business and technology.

Firms can achieve successful RFID implementations that add value given the determination to rise above unexpected challenges. Knowledge, skill, and leadership will play a major role in successful RFID implementations using the EPCglobal Network. In the words of Edward Gibbon, "the winds and the waves are always on the side of the ablest navigators." [33]

---

[e] http://www.w3.org/

Notes

Preface

1. Gibbon, Edward (1910), *The Decline and Fall of the Roman Empire – Volume 6*, New York: Alfred A.Knopf, p. 619.

2. Lord Byron (1808), "Inscription Honoring Boatswain," Newstead Abby: England, November 30.

Chapter 1 – The Emergence of a New Key Technology

1. Haberman, Alan L., Editor (2001), *Twenty-Five Years Behind Bars*, Cambridge, MA: Harvard University Press, p. 2.

2. Pommer, Michael D. and Eric N. Berkowitz (1980), "UPC Scanning: An Assessment of Shopper Response to Technological Change," *Journal of Retailing* 56:2, pp 27 – 28.

3. Coyle, John J., Edward J. Bardi and C. John Langley (1992), *The Management of Business Logistics*, New York: West Publishing Company.

4. Simchi-Levi, David, Philip Kaminsky and Edith Simchi-Levi (2002), *Designing and Managing the Supply Chain*, New York: McGraw-Hill/Irwin.

5. Ibid, ref. 1, p. 28.

6. Fergusson, Nicholas (2005), Princeton, NJ: EPCglobal, Inc., personal correspondence.

7. Ibid, ref. 1, p. 27.

8. Fensel, Dieter, James Hendler, Henry Lieberman, and Wolfgang Wahlster (2003), *Spinning the Semantic Web*, Cambridge, MA: MIT Press, p. 363.

9. Sarma, Sanjay, David L. Brock, and Kevin Ashton (2000), "The Networked Physical World," Cambridge, MA: *MIT Auto-ID Center*.

10. Scharfeld, Tom Ahlkvist (2001), "An analysis of the Fundamental Constraints on Low Cost Passive Radio-Frequency Identification System Design," Unpublished Thesis (Master of Science), Cambridge, MA: *Massachusetts Institute of Technology*.

11. Shah, Ronak (2005), "RFID Without the Hype: Best Practices and Lessons Learned," LFM Alumni Webinar, Engineering Systems Division, Cambridge, MA: *Massachusetts Insitute of Technology*, May 20.

12. Takahashi, Dean (2004), "The Father of RFID," *San Jose Mercury News*, June 7.

13. Schoenberger, Chana R. (2002), "The Internet of Things," *Forbes*, March 18.

14. Roberti, Mark (2005), "The Mood of the EPCglobal Community," *RFID Journal*, September 19.

15. Matlick, Justin (2005), "It's 10 a.m. at the Water Park: Do You Know Where Your Children Are?" *The Wall Street Journal*, April 25.

16. Ward, Sandra (2004), "Making Waves," *Barron's*, May 11.

17. Sullivan, Laurie (2005), "RFID is here to Stay," *InformationWeek*, March 7.

18. Colvin, Geoff (2004), "Voice over IP," *Wall Street Week*, July 16.

For a transcript of the program, go to the following link:
http://www.pbs.org/wsw/tvprogram/20040716.html

Note: This observation is also attributed to Alan Abelson of Barron's.

19. Greenspan, Alan (2005), "On Economic Flexibility," Chicago, IL: *National Association for Business Economics Annual Meeting*, September 27.

20. Koh, Robin, Edmund W. Schuster, Indy Chackrabarti, and Attilio Bellman (2003), "Securing the Pharmaceutical Supply Chain," Cambridge, MA: *The MIT Auto- ID Center*.

21. Schuster, Edmund W. and Robin Koh (2004), "To Track and Trace," *APICS – The Performance Advantage* 14:2, pp. 34 – 38.

22. Koh, Robin, Edmund W. Schuster, Nhat-So Lam, and Mark Dinning (2003), "Prediction, Detection, and Proof: An Integrated Auto-ID Solution to Retail Theft," Cambridge, MA: *The MIT Auto-ID Center*.

23. Kar, Pinaki, Ming Li, and Edmund W. Schuster (2003), "A Case Study of Computer Service Parts Inventory Management," Service Parts Resource Guide (1st Edition), Falls Church, VA: *American Production and Inventory Control Society*.

24. Gilliam, Tig and Herb Kleinberger (2003), "Business Case Research Track," Cambridge, MA: *MIT Auto-ID Center*, February 26.

25. Engels Daniel W., Robin Koh, Elaine Lai, and Edmund W. Schuster (2004), "Improving Visibility in the DOD Supply Chain," *Army Logistician* 36:3, pp. 20 – 23.

26. Hays, Constance L. (2004), "What Wal-Mart Knows About Customers' Habits," *The New York Times*, November 14.

27. Cox, Braden and Fred L. Smith, Jr. (2005), "Slicing Telecom the Right Way," *Barron's*, July, 25.

28. Buchanan, Mark (2002), *Small Worlds and the Groundbreaking Theory of Networks*, New York, NY: W.W. Norton, p. 77.

29. Drucker, Peter (1999), "Beyond the Information Revolution," *Atlantic Monthly* 284 (October), pp. 47-57.

30. Gibbon, Edward (1910), *The Decline and Fall of the Roman Empire – Volume 4*, New York: Alfred A.Knopf, p. 449.

31. Leyden, Peter and Peter Schwartz (1997), "The Long Boom," *Barron's*, September 1.

32. Donlan, Thomas G. (2003), "Following the Wrights," *Barron's*, December 15.

33. Anderson, Sherwood (1920), *Poor White*, New Directions: New York, p. 16.

34. Ibid, ref 1, p. 28.

35. Oliner, Stephen D. and Daniel E. Sichel (2000), "The Resurgence of Growth in the Late 1990s: Is Information Technology the Story?" *The Federal Reserve Board: Finance and Economics Discussion Series*, May 17.

36. Ibid, p. 2.

37. Sieling, M., B. Friedman, and M. Dumas (2001), "Labor Productivity in the Retail Trade Industry, 1987–99," *Monthly Labor Review* 124:12, pp. 3–14.

38. Wessel, David (2005), Better Information Isn't Always Beneficial," *The Wall Street Journal*, September 22.

39. Ibid, ref. 19.

Chapter 2 – Hardware: RFID Tags and Readers

1. Dinning, Mark and Schuster Edmund W. (2003), "Fighting Friction," *APICS – The Performance Advantage* 13:2, pp. 27 – 31.

2. Engels, Daniel W., Sanjay E. Sarma, Laxmiprasad Putta, and David Brock (2002), "The Networked Physical World System," *Proceedings of the IADIS International Conference on WWW/Internet* 2002: Lisbon, Portugal, November 13-15.

3. Ibid.

4. Cole, Peter (2002), A Study of Factors Affecting the Design of EPC Antennas & Readers for Supermarket Shelves," Cambridge, MA: *MIT Auto-ID Center*.

5. Haberman, Alan – General Editor (2001), *Twenty-Five Years Behind Bars*, Cambridge, MA: Harvard University Press.

6. Unauthored (2005), "Active RFID – A profitable business," *IDTechEx*, Dec 12.

7. Sarma, Sanjay (2001), "Towards the 5¢ Tag," Cambridge, MA: *MIT Auto-ID Center*.

8. Collins, Jonathan (2005), "Rafsec Announces Gen 2 Pricing," *RFID Journal*, September 20.

9. Ibid, ref 7.

10. Ibid.

11. Ibid.

12. Kumar, Arun (2005), "RFID in India," *RFID Journal*, December 4.

13. Unauthored, "RFID System Components and Costs," *RFID Journal*.

14. Dunlap, Joe, Greg Gilbert, Lyle Ginsburg, Paul Schmidt, and Jeff Smith (2003), "If You Build It, They Will Come: EPC Forum Market Sizing Analysis," Cambridge, MA: *MIT Auto-ID Center*.

15. Ibid, ref. 6.

16. Ibid. ref. 14.

17. Unauthored (2003), Tyco to Mass-Produce RFID Readers, *RFID Journal*, February 24.

18. Klym, Natalie, Charlie Fine, Dirk Trossen, and Milind Tavshikar (2006), "The Evolution of RFID Networks: The Potential for Disruptive Innovation," Cambridge, MA: *MIT Communications Futures Program*.

19. Schlesinger, Adam Ian (2005), "Mitigating Container Security Risk Using Real-Time Monitoring with Active Radio Frequency Identification and Sensors," Unpublished Thesis (Master of Engineering), Cambridge, MA: *Massachusetts Institute of Technology:Cambridge*.

20. Brock, David L. (2001), "The Electronic Product Code – A Naming Scheme for Physical Objects," Cambridge, MA: *MIT Auto-ID Center*.

21. Brock, David L (2001), "The Compact Electronic Product Code – A 64-bit Representation of the Electronic Product Code," Cambridge, MA: *MIT Auto-ID Center*.

22. Ibid, ref, 2.

23. Schuster, Edmund W., Tom A. Scharfeld, Pinaki Kar, David L. Brock, and Stuart J. Allen (2004), "The Next Frontier: How Auto-ID could Improve ERP Data Quality," *Cutter IT Journal* 17:9.

24. Scharfeld, Tom A. (2003), "Compliance and Certification: Ensuring RFID Interoperability," Cambridge, MA: *MIT Auto-ID Center*.

25. Ibid

26. Collins, Jonathan (2004), "Report Weighs Impact of EPC," *RFID Journal*, October 14.

27. Sliwa, Carol (2005), "RFID Tag Price Must Fall for Users to Reap Rewards," *Computerworld*, April 27.

28. Sullivan, Laurie (2004), "UPDATE: Next-Generation RFID Standard Is Approved," *RFID Journal*, December 17.

29. Wolf Bielas (2005), "Moving Toward Gen 2 RFID," *RFID Journal*, September 26.

30. Roberti, Mark (2004), "The Role of Independent Testing," *RFID Journal*, September 27.

31. Fletcher, Rich (2004), "Packaging and RFID Special Interest Group," Cambridge, MA: *MIT Auto-ID Labs*, March 10.

32. Marti, Uttara P. (2005), "Enhancement of Electromagnetic Propagation Through Complex Media for Radio Frequency Identification," Unpublished Thesis (Master of Engineering), Cambridge, MA: *Massachusetts Institute of Technology*.

33. Wolk, Jonathan E. (2005), "Graphical Real-Time Simulation Tool for Passive UHF RFID Environments," Unpublished Thesis (Master of Engineering), Cambridge, MA: *Massachusetts Institute of Technology*.

34. Redemske, Richard Michael (2005), "An Electromagnetic Measurement Tool for UHF RFID Diagnostics," Unpublished thesis (Master of Engineering), Cambridge, MA: *Massachusetts Institute of Technology*.

35. Scharfeld, Tom A. (2001), "An Analysis of the Fundamental Constraints on Low Cost Passive Radio-Frequency Identification system Design," Unpublished Thesis (Master of Science), Cambridge, MA: *Massachusetts Institute of Technology*.

36. Ukkonen, Leena, Mikael Soini, Daniel W. Engels, Lauri Sydanheimo, and Marku Kivikoski (2004), "Effect of Conductive Material in Objects on Identification with Passive RFID Technology: a Case Study of Cigarette Cartons," *ICMA International Conference on Machine Automation*: Osaka, Japan, November 24-26.

37. Ukkonen, Leena, Daniel W. Engels, Lauri Sydanheimo, and Marku Kivikoski (2005), "Folded Microstrip Patch Antenna for RFID Tagging of Objects Containing Metallic Foil," *Proceedings of the 2005 IEEE AP-S International Symposium*: Washington D.C., July 5-8.

38. Delichatsios, Stefanie Alki, and Daniel W. Engels (2006), "The Albano Passive UHF RFID Tag Antenna," *Proceedings of Waveform Diversity & Design Conference*: Lihue, Hawaii, January 22-27.

39. O'Connor, Mary Catherine (2005), "New System for Printing Tags," *RFID Journal*, September 21.

Chapter 3 – Infrastructure: EPCglobal Network

1. Kay, Marshall (2003), "EPC and the Need for Good Data," *RFID Journal*, August 4.

2. Fine, Charles H., Natalie Klym, and Dirk Trossen (2005), RFID: Exploring Edges and Potential for Disruption, published in *RFID and Beyond: Growing Your Business through Real World Awareness*, Claus Heinrich (editor), New York: John Wiley & Sons.

3. Engels, Daniel W., Sanjay E. Sarma, Laxmiprasad Putta, and David Brock (2002), "The Networked Physical World System," *Proceedings of the IADIS International Conference on WWW/Internet* 2002: Lisbon, Portugal, November 13-15.

4. Dinning, Mark and Edmund W. Schuster (2003), "Fighting Friction," *APICS – The Performance Advantage* 13:2.

5. Brock, David L. (2001), "The Physical Markup Language – A Universal Language for Physical Objects," Cambridge, MA: *MIT Auto-ID Center*.

6. Ibid, ref. 3.

7. Ibid.

8. Hogan, Bernie (2006), "EPCglobal Update," RFID Convocation, Cambridge MA: *Massachusetts Institute of Technology*, January 23.

9. Ibid.

10. Roberti, Mark (2005), "The Mood of the EPCglobal Community," RFID Journal, September 19.

11. Ibid.

12. "End to End Visibility," *RFID Journal Bulletin Board*. Posted on 06/14/2004.

13. Ibid.

14. Ibid

15. Inaba, Tatsuya (2004), "An Analysis of Physical Object Information Flow within Auto-ID Infrastructure," Unpublished Thesis (Master of Engineering), Cambridge, MA: *Massachusetts Institute of Technology*, p. 2.

16. Ibid, ref. 10.

17. O'Connor, Mary Catherine (2005), "Startup Opens Up RFID Middleware," *RFID Journal*, April 6, 2005.

18. Ibid.

19. Ibid.

20. O'Connor, Mary Catherine (2004), UCLA Consortium Holds RFID Forum," *RFID Journal*, October 18.

21. Sullivan, Laurie (2004), "UCLA Develops RFID Middleware," *InformationWeek*, October 1.

22. Roberti, Mark (2005), "Lab to Build EPC Network Simulation," *RFID Journal*, August 19.

23. Ibid.

24. Baum, Mark 2005), "Steady Progress Wins the Race for All," *RFID Journal*, April 4.

25. Brock, David L. (1993), "A Task Level Control System for Robotic Grasping," Unpublished Thesis (Doctor of Philosophy), Cambridge, MA: *Massachusetts Institute of Technology*.

26. Brock, David L. (1992), "Remote Controlled Machinery, Processes over the Web Closer than You Think," *The MIT Report*, September.

27. Metcalfe, Robert (1992), "XCoffee," *Communications Week*, January 1992.

28. Ibid, ref. 26

29. O'Brien, Kevin M. (1996), "Task-level Control for Networked Telerobotics," Unpublished Thesis (Master of Science), Cambridge, MA: *Massachusetts Institute of Technology*.M.S.

30. Tokgoz, Ozgu (1998), "A Simulation Environment for Multi-User Telerobotics," Unpublished Thesis (Master of Science), Cambridge, MA: *Massachusetts Institute of Technology*.

31. Lomigora, Nedzad (1999), "Cost-effective Approach to Automated Resource Management using Real-time Sensing and Networking," Unpublished Thesis (Master of Science), Cambridge, MA: *Massachusetts Institute of Technology*.

32. Garfinkel, Simson L. and Beth Rosenberg (2005), *RFID: Applications, Security, and Privacy*, Boston, MA: Addison – Wesley.

Chapter 4 – Data: What, When, and Where?

1. Schuster, Edmund, W., David L. Brock, Stuart J. Allen, Pinaki Kar, and Mark Dinning (2005), "Enabling ERP through Auto-ID Technology." In *Strategic ERP Extension and Use*, edited by E. Bendoly and F.R. Jacobs. Palo Alto: Stanford University Press.

2. Gleick, James (2003), *Isaac Newton*, New York: Pantheon Books, p. 74 – 75.

3. Mathews, Anna Wilde (2004), "FDA Plans Bogus-Drug Crackdown," *The Wall Street Journal*, February 19.

4. Carr, David .F. and Larry Barrett (2005), "Philip Morris International: Smoke Screen," *Baseline, The Project Management Center,* February 1.

5. Becker, Christian and Frank Duerr (2005), "On Location Models for Personal, and Ubiquitous Computing," *Personal and Ubiquitous Computing,* 9:1, pp. 20 – 31.

Chapter 5 – Warehousing: Improving Customer Service

1. Fine, Charles H. (1998), *Clockspeed*, Reading, MA: Perseus Books, p. 30.

2. Ibid, p. 31.

3. Coyle, John J., Edward J. Bardi and C. John Langley, Jr (1992), *The Management of Business Logistics*, New York: West Publishing Company, p. 347.

4. Unauthored (2006), "Everyone Likes to Laud Serving the Customer; Doing It Is the Problem," *The Wall Street Journal*, February 27.

5. Hudson, Kris (2005), "Wal-Mart's Need for Speed," *The Wall Street Journal*, September 26.

6. Dinning, Mark (2002), "Applications of Auto-ID Technology to Gain Supply Chain Process Efficiencies in the Consumer Packaged Goods Industry," Unpublished Thesis (Master of Engineering), Cambridge, MA: *Massachusetts Institute of Technology*.

7. Ibid.

8. Sterling, Jay U. and Douglas M. Lambert (1987), "Establishing Customer Service Strategies within the Marketing Mix," *Journal of Business Logistics* 8:1.

9. Innis, Daniel E. and Bernard J. LaLonde (1994), "Customer Service: the Key to Customer Satisfaction, Customer Loyalty, and Market Share," Journal of Business Logistics, 15:1.

10. Ibid.

11. Ibid, ref. 8.

12. Ibid, ref. 9.

13. Ibid, ref. 8.

14. Ibid, ref. 3, p. 83.

15. Ibid, ref. 8.

16. Hechinger, John and Laura Johannes (2000), "Xerox Considers a Major Revamping, Suffering from Self-Inflicted Wounds," *The Wall Street Journal*, October 20.

17. Alexander, Keith, Tig Gilliam, Kathryn Gramling, Mike Kindy, Dhaval Moogimane, Mike Schultz, Maurice Woods (2002), "Focus on the Supply Chain: Applying Auto-ID within the Distribution Center," Cambridge, MA: *MIT Auto-ID Center*.

18. Ibid, ref, 6.

19. Ibid, ref 17.

20. Dinning, Mark and Edmund W. Schuster (2003), "Fighting Friction," *APICS – The Performance Advantage* 13:2.

21. Ibid, ref, 6.

22. Ibid

23. Ibid.

24. Laing, Jonathan R. (2000), "The New Math," *Barron's*, November 20.

25. Bernasek, A (2002), "The Friction Economy: American Business Just Got the Bill for the Terrorists Attack: $151 billion – a year," *Fortune Magazine*, February 18.

Chapter 6 – Maintenance: Service Parts Inventory Management

1. Srinivasan, Krishnan (2002), "Service Parts Logistics at Caterpillar: What Works – Both Old and New," Service Parts Inventory Management, Center for Transportation and Logistics: Cambridge, MA: *Massachusetts Institute of Technology*, March 20.

2. Tibben-Lembke, Ronald S. and Henry N. Amato (2001), "Replacement Parts Management: The Value of Information," *Journal of Business Logistics* 22:2, p. 149.

3. Ibid, p. 163.

4. Kar, Pinaki, Ming Li, and Edmund W. Schuster (2003), "A Case Study of Computer Service Parts Inventory Management," Service Parts Resource Guide, 1st Ed. Falls Church, VA: *American Production and Inventory Control Society*.

5. Cohen, Morris A. and Vipul Agarwal (1999), "After-Sales Service Supply Chains: A benchmark Update of the North American Computer Industry," Fishman-Davidson Center for Service and Operations Management, Philadelphia, PA: *University of Pennsylvania*.

6. Ibid.

7. Bazovsky, Igor (1961), *Reliability Theory and Practice*, Englewood Cliffs, NJ: Prentice-Hall Inc.

8. Graves, S.C. (1985), "A Multi-Echelon Inventory Model for a Repairable Item with One-for-One Replenishment," *Management Science* 31:10, pp. 1247-1256.

9. Hillestad, Richard J. and Manuel J. Carrillo (1980), "Models and Techniques for Recoverable Item Stockage When Demand and the Repair Processes are Non-Stationary – Part I : Performance Measurement," Santa Monica, CA: *RAND Corporation*.

10. Sherbrooke, Craig C. (1992), *Optimal Inventory Modeling of Systems: Multi-Echelon Techniques*, New York: John Wiley and Sons Inc.

11. Schuster, Edmund W. (2004), Personal correspondence.

12. O'Connor, Mary Catherine (2005), "Boeing's Flight Plan for Dreamliner Tags," *RFID Journal*, October 28.

13. Parlikad, Ajith Kumar, Duncan McFarlane, Elgar Fleisch, and Sandra Gross (2003), "The Role of Product Identity in End-of-Life Decision Making," Cambridge, MA: *MIT Auto-ID Center*.

Chapter 7 – Pharmaceuticals: Preventing Counterfeits

1. Cottrill. Ken (2001), "Blockbuster Market," *Traffic World* 265:27.

2. Mitchell, P. (1998), "Documentation: an Essential Precursor to Drug Manufacturing," *APICS – The Performance Advantage*, September.

3. FDA (2004), "Combating Counterfeit Drugs," *Food and Drug Administration* February 19.

4. Whiting, Rick (2004), "RFID Could Be Boon To Small Businesses," *InformationWeek*, May 19.

5. Malykhina, Elena (2004), "RFID Tests Are Positive For CVS And Pharmaceuticals," *InformationWeek*, Sept. 30.

6. Brock, David L. (2002), "Smart Medicine – The Application of Auto-ID Technology to Healthcare," Cambridge, MA: *MIT Auto-ID Center*.

7. World Health Organization (July 10, 2005).
http://www.wpro.who.int/media_centre/fact_sheets/fs_20050506.htm

8. *Healthcare Distribution Management Association* (2003), "Pharmaceutical Product Tampering News Media Factsheet," March.

9. Newton, Paul, Stephane Proux, Michael Green, Frank Smithuis, Jan Rozendaal, Sompol Prakongpan, Kesinee Chotivanich, Mayfong Mayxay, Sornchai Looareesuwan, Jeremy Farrar, Francois Nosten, and Nicholas J. White (2001), "Fake Artesunate in Southeast Asia," *The Lancet* 357, pp. 1948 – 1949.

10. Goodman, Peter S. (2002), "China's Killer Headache: Fake Pharmaceuticals," *The Washington Post*, August 30.

11. Pasternak, Douglas (2001), Knockoffs on the Pharmacy Shelf," *U.S. News & World Report* 130:23.

12. Taylor, R.B., O Shakoor, R.H. Behrens, M Everard, A.S. Low, J. Wangboonskul, R.G. Reid, J.A. Kolawole (2001), "Pharmacopoeial Quality of Drugs Supplied by Nigerian Pharmacies," *The Lancet* 357, pp. 1933 – 1936.

13. Ibid, ref. 11.

14. Appleby, Julie (2003), "Fake Drugs Show Up in U.S. Pharmacies; As Prescription Prices Rise, Counterfeiters Chase Profits," *USA TODAY*, May 15.

15. Ibid.

16. Saul, Stephanie (2005), "Making a Fortune by Wagering that Drug Prices Tend to Rise," *The New York Times*, January 26.

17. Ibid, ref. 14.

18. Florida Prescription Drug Protection Act Update
http://www.doh.state.fl.us/pharmacy/drugs/news/SB2312Effdates.pdf

19. Chackrabarti, I. (2003), "An Auto-ID Based Approach to Reduce Counterfeiting in the U.S.Pharmaceutical Supply Chain," Unpublished Thesis (Master of Engineering), Cambridge: MA, *Massachusetts Institute of Technology*.

20. Ibid, ref 16.

21. http://www.supplyscape.com/industries/

22. Bellman, Attilio (2003), "Product Traceability in the Pharmaceutical Supply Chain: An Analysis of the Auto-ID Approach," Unpublished Thesis (Master of Engineering), Cambridge, MA: *Massachusetts Institute of Technology*.

23. Ibid, ref 11.

24. (www.fda.gov/medwatch/index.html)

25. (www.healthcaredistribution.org)

26. (www.ismp.org)

27. (www.productsurety.org)

28. Milne, T.P. (2002), "Auto-ID Business Use-Case Framework (A-Biz) – Background," Cambridge, MA: *MIT Auto-ID Center*.

29. Harrison, Mark G., Humberto J. Moran, James P. Brusey, and Duncan C. McFarlane (2003), "PML Server Developments," Cambridge, MA: *MIT Auto-ID Center*.

30. Milne, T.P. (2002), "Auto-ID Business Use-Case Framework (A-Biz) – Despatch Advice Use-Case," Cambridge, MA: *MIT Auto-ID Center*.

31. Ibid, ref. 29.

32. For examples, see http://www.gs1.org/productssolutions/gdsn/

33. Ibid, ref. 29.

34. Mathews, Anna Wilde (2004), "FDA Plans Bogus-Drug Crackdown," *The Wall Street Journal*, February 19.

35. Greengard, Samual (2003), "RFID: Cure for Counterfeit Drugs?" *RFID Journal*, October 15.

36. Select Excerpt from the FDA web site:

http://www.governmentguide.com/govsite.adp?bread=*Main&url=http%3A//www.governmentguide.com/ams/clickThruRedirect.adp%3F55076483%2C16920155%2Chttp%3A//www.fda.gov

37. Philipkoski, Kristen. (2003), "Designer Drugs: Fact or Fiction?" *Wired News*, January 17.

38. Arnst, Catherine, Amy Barrett, Michael Arndt, and John Carey (2004), "The Wanning of the Blockbuster Drug," *BusinessWeek*, October 18.

39. Alpert, Bill (2005), "Roche's Revolution," *Barron's*, April 4.

40. Ibid, ref. 38.

41. Ibid, ref. 39.

42. Ibid, ref. 39

43. Tesoriero, Heather Won (2005), "Cardinal Health Ends Drug Trading," *The Wall Street Journal*, May 6.

44. Eban, Katherine (2005), *Dangerous Doses: How Counterfeiters Are Contaminating America's Drug Supply*, New York: Harcourt.

45. Balfour Frederik, Carol Matlack, Amy Barrett, Kerry Capell, Dexter Roberts, Jonathan Wheatley, William C. Symonds, Paul Magnusson, and Diane Brady (2005), "Fakes," *BusinessWeek*, February 7.

46. Christian, James (2001), "Continuing Concerns over Imported Pharmaceuticals," The Committee on Energy and Commerce (W.J. Tauzin – Chairman), Subcommitte on Oversight and Investigations, *107th Congress House Hearings*, pp. 157 – 173. http://energycommerce.house.gov/107/Hearings/06072001hearing267/print.htm

47. Tesoriero, Heather Won (2005), "Microsoft Joins Pfizer to Fight 'Viagra' Spam," *The Wall Street Journal*, February 11.

48. Staake, Thorsten, Frédéric Thiesse and Elgar Fleisch (2005)," Extending the EPC Network – The Potential of RFID in Anti-Counterfeiting," *Symposium on Applied Computing Proceedings of the 2005 ACM Symposium on Applied Computing – Ubiquitous Computing Session*: Santa Fe, New Mexico, pp. 1607 – 1612.

49. Bowe, Christopher (2005), "Pharma Giants Under Pressure to Cut Costs," *Financial Times*, June 20.

50. Hume, David (1778), *The History of England*, Indianapolis, Indiana: Liberty Fund, vol. 3, p. 74.

This book is available free of charge at http://www.libertyfund.org/. The site contains downloadable pfds of the complete book.

51. Beier, Frederick J. (1995), "The Management of the Supply Chain for Hospital Pharmacies: A Focus on Inventory Management Practices," *Journal of Business Logistics* 16:2.

52. Taylor, Sam G., S.M. Seward and Steve F. Bolander (1981), "Why the Process Industries are Different," *Production and Inventory Management Journal* 22:4, pp. 9-24.

53. Schuster, Edmund W., Stuart J. Allen, and Michael P. D'Itri (2000), "Capacitated MRP and its Application in the Process Industries," *Journal of Business Logistics* 21:1, pp. 169 – 189.

54. Umble, M.M. (1992), "Analyzing Manufacturing Problems Using V-A-T Analysis," *Production and Inventory Management Journal* 33:2, p. 55-60.

55. Goodspeed, Peter (2001), "'No Way of Knowing [if] He Gave Me the Right Doses': Kansas City Arrest: Sample Had Potency of Less Than 1%, FBI Test Shows," *National Post*, August 17.

56. Harris, Gardiner (2004), "Tiny Antennas to Keep Tabs on U.S. Drugs," *The New York Times*, November 15.

57. Ibid.

58. McKesson, annual revenue for 2004, $57.1 billion – Rank 16
Cardinal Health, annual revenue for 2004, $56.8 billion – Rank 17
AmerisourceBergen, annual revenue for 2004, $49.7 billion – Rank 22
Source: Fortune Magazine
http://www.fortune.com/fortune/subs/500archive/company/0,19388,A,00.html

59. Ibid.

60. Davies, Paul and Joann Lublin (2005), Bristol CEO Relies on Board Allies," *The Wall Street Journal*, July, 1.

61. Saul, Stephanie (2005), "Making a Fortune by Wagering that Drug Prices Tend to Rise," *The New York Times*, January, 26.

62. Ibid, ref. 56.

63. Tesoriero, Heather and Gary Fields (2003), "FBI, FDA Investigate Big Drug Wholesaler," *The Wall Street Journal*, September 19.

64. Koh, Robin., Edmund W. Schuster, Indy Chackrabarti, and Attilio Bellman (2003), "Securing the Pharmaceutical Supply Chain," Cambridge, MA: *MIT Auto-ID Center*.

65. Voyles, Bennett (2005), "Small Tech Firm Seeks Fees From Pharma," *RFID Journal*, June 22.

66. Albano, Silvio and Daniel Engels (2001), "Auto-ID Center Field Trial: Phase I Summary," Cambridge, MA: *MIT Auto-ID Center*.

67. Milne, Timothy, P. (2002), "Auto-ID Business Use-Case Framework (A-Biz) – Background," Cambridge, MA, *MIT Auto-ID Center*.

68. Whiting, Rick (2004), "RFID Could Be Boon To Small Businesses," *InformationWeek*, May 19.

69. Whiting, Rick (2004), "RFID to Flourish in Pharmaceutical Industry," *InformationWeek*, Aug. 23.

70. EPCGlobal, http://www.epcglobalinc.org/

71. Voyles, Bennett (2005), "Group Finalizes Drug Security Network," *RFID Journal*, May 9.

72. Scharfeld, Tom A. (2004), "Introduction to HRI Research: Frequency and Safety," Cambridge, MA: *MIT Auto-ID Labs*.

73. Ibid.

74. Inaba, Tatsuya and Edmund W. Schuster (2005), "Meeting the FDA's Initiative for Protecting the US Drug Supply," Submitted for publication in *American Pharmaceutical Outsourcing Journal*.

75. Nickle, Bill (2004), "Pitfalls and Practical Solutions for RFID," *RFID Workshop*: Rutgers University, November 15.

76. Engels, Daniel W. (2004), "MIT Healthcare Research Initiative: Vision, Challenges, and Research," Cambridge, MA: *MIT Auto-ID Labs*, December 9.

77. Klein, A.A. and G.N. Diaiani (2003), "Mobile Phones in the Hospital – Past, Present and Future," *Anaesthesia* 58:4, p. 353.

78. Sarma, Sanjay E. (2001), "Towards the 5c Tag," Cambridge, MA: *MIT Auto-ID Center*.

79. Ward, Diane Marie (2004), "5-Cent Tag Unlikely in 4 Years," *RFID Journal*, August 26.

80. Tesoriero, Heather Won (2004), "Radio ID Tags will Help Monitor Drug Supplies," *The Wall Street Journal*, November 16.

81. Malykhina, Elena (2004), "Bar Codes Expected to Have a Long Life," *InformationWeek*, October 21.

82. Ibid, ref. 29.

83. Personal correspondence, Dave Chanoux – Scanning Devices Inc.

84. O'Connor, Mary Catherine (2005), "TI, VeriSign Devise Drug-Protection Plan," *RFID Journal*, June 1.

85. O'Connor, Mary Catherine (2005), "Bar Coding for Item Tracking," *RFID Journal*, January 6.

86. Malykhina, Elena (2004), "The Case for Bar Codes," *InformationWeek*, December 3.

87. Ibid, ref. 85.

88. Tesoriero, Heather Won (2005), "Cardinal Health Ends Drug Trading," *The Wall Street Journal*, May 6.

89. Tesoriero, Heather Wan (2005), "CVS Will Try to Close One Gap Where Counterfeit Drugs Enter," *The Wall Street Journal*, May 25.

90. Ibid, ref. 61.

91. Ibid.

92. Ibid.

93. Solomon, Jay and Gordon Fairclough (2005), "North Korea's Counterfeit Goods Targeted," *The Wall Street Journal*, June 1.

94. Apuzzo, Matt (2005), "Subway Torpedoes 'Sub Club' Promotion Sandwich Maker Cites Stamp Counterfeiting," *The Associated Press*, June 3.

95. Ibid, ref. 36.

Chapter 8 – Medical Devices: Smart Healthcare Infrastructure

1. Unauthored (2005), "From Intel to Health Care and Beyond," *The New York Times*, July 30.

2. Tesoriero, Heather Won (2005), "Medical Spending Rose 8.2% in '04, Down From '03," *The Wall Street Journal*, June 2005.

3. Unauthored (2005), "Greenspan Sees Medicare Crunch," *The Wall Street Journal*, February 17.

4. Unauthored (2005), "RFID Use in Healthcare Set to Take Off," *RFID Journal*, April 26.

5. Unauthored (2005), "Poll Indicates Strong Support for New Medical Technologies," *The Wall Street Journal*, October 7.

6. Whalen, Jeanne (2005), "Tailored Medicine," *The Wall Street Journal*, March 18.

7. Ingebretsen, Mark (2004), "RFID Technology Could Have Many Health-Care Applications, *The Wall Street Journal*, July 14.

8. Gramling, Kathryn, Anthony Bigornia and Tig Gilliam (2003), "EPC Forum Survey," Cambridge, MA: *The MIT Auto-ID Center*.

9. O'Connor, Mary Catherine (2005), "Sun Debuts Solutions for Asset Tracking," *RFID Journal*, October 27.

10. Alvarez, Gene (2004), "RFID Helps Enterprises Increase Return On Assets Through Tracking," *InformationWeek*, December 21.

11. Ibid.

12. Ibid, ref. 7.

13. Unauthored (2005), "Duke Says Patients Won't Face Infection After Mishap With Surgical Instruments," *The Wall Street Journal*, June 21.

14. Ibid.

15. Unauthored (2004), "FDA Clears Implantable Chip That Can Provide Patient Data," *The Wall Street Journal*, October 13.

16. Dooren, Jennifer Corbett (2005), "FDA Warns of Drug Confusion," *The Wall Street Journal*, February 9.

17. Ibid.

18. Ibid.

19. Cai, Qingyun, Kefeng Zeng, Chuanmin Ruan, Tejal A. Desai, and Craig A. Grimes (2004), "A Wireless, Remote Query Glucose Biosensor Based on a pH-Sensitive Polymer," *Analytical Chemistry*; 76:14.

20. Mun, In K. (2004), "RFID for Hospitals," MIT Healthcare Research Innitiative, Cambridge, MA: *Massachusetts Institute of Technology*, December 9.

21. Malyhina, Elena (2004), Bar Codes Expected to have a Long Life," *Information-Week*, October 21.

22. Ibid, ref. 20.

23. Ibid, ref. 5.

24. Unauthored (2005), "Five Innovations Aid the Push To Electronic Medical Records," *The Wall Street Journal*, February 9.

25. Ibid.

26. Langlois, Richard (1997), "Standards, Modularity, and Innovation: the Case of Medical Practice," *Proceedings of the Conference on Path Dependence and Path Creation*.

Chapter 9 – Agriculture: Animal Tracking

1. Antle, John M. (1996), "Efficient Food Safety Regulation in the Food Manufacturing Sector," *American Journal of Agricultural Economics* 78:6, pp. 1242 – 1247.

2. Banwart, George J. (1979), *Basic Food Microbiology*, Westport, Connecticut: AVI Publishing Company, Inc., p. 221.

3. Thomsen, Michael R., Andrew M. Mckenzie (2001), "Market Incentives for Safe Foods: An Examination of Shareholder Losses from Meat and Poultry Recalls," *American Journal of Agricultural Economics* 82:3, pp. 526 – 538.

4. Hobbs, Jill E. (2004), "Information Asymmetry and the Role of Traceability Systems" *Agribusiness* 20:4, pp 347 – 415.

5. Ibid.

6. Ibid.

7. Unnevehr, Laurian (2004), "Mad Cows and Bt Potatoes: Global Public Goods in the Food System," *American Journal of Agricultural Economics* 86:5, pp 1159 – 1166.

8. Salin, Victoria (2000), "Information Technology and Cattle-Beef Supply Chains," *American Journal of Agricultural Economics* 82:5, pp 1105 – 1111.

9. Ibid.

10. Mathews, Jr. Kenneth H., William F. Hahn, Kenneth E. Nelson, Lawrence A. Duewer, and Ronald A. Gustafson (1999), U.S. Beef Industry: Cattle Cycles, Price Spreads, and Packer Concentration," Technical Bulletin No. (TB1874), Washington, D.C.: *U.S. Department of Agriculture*.

11. Kilman, Scott (2003), "Farm Belt Becomes Driver for the Overall Economy," *The Wall Street Journal*, December 17.

12. Kilman, Scott, Shirley Leung, and Tamsin Carlisle (2003), "Case of Mad Cow Found in the U.S. for First Time," *The Wall Street Journal*, December 24.

13. Stecklow, Steve and Scott Kilman (2004), "As U.S. Pleads Mad-Cow Case, Past Practices Are a Handicap," *The Wall Street Journal*, January 8.

14. Ibid.

15. Carlisle, Tamsin (2004), "Canada Says Farm Sector Lost Money in 2003," *The Wall Street Journal*, February 8.

16. Ibid. ref 12.

17. Regalado, Antonio (2003), "Scientific Data Offer No Proof Of Beef Safety," *The Wall Street Journal*, December 29.

18. Kilman, Scott (2004), "U.S. Concludes Mad-Cow Investigation," *The Wall Street Journal*, February 10.

19. Wald, Matthew (2003), U.S. Scours Files to Trace Source of Mad Cow Case, The Wall Street Journal, December 25.

20. Ibid.

21. Hanson, Eric (2004), "Swift Launches Traceability Program," *Meatingplace*, January 2.

22. Unauthored (2005), "USDA Proposes System To Track Individual Cattle," *Associated Press*, May 6.

23. Brat, Ilan (2006), "New Kind of Cattle Branding," *The Wall Street Journal*, March 1.

24. Swedberg, Claire (2005), "Cattle Auctioneer Promotes Tracking Plan," *RFID Journal*, June 13.

25. Ibid.

26. Mun Leng Ng, Kin Seong Leong, David M. Hall, Peter H. Cole (2005), "A Small Passive UHF RFID Tagfor Livestock Identification," *Proceedings of IEEE 2005 International Symposium on Microwave, Antenna, Propagation and EMC Technologies For Wireless Communications*: Beijing, China, August 8-12, 2005

Chapter 10 – Food: Dynamic Expiration Dates

1. Potter, Norman N. (1973), *Food Science*, Westport, CT: The AVI Publishing Company, p. 132.

2. Ibid, p. 133.

3. Welch Foods, Inc., Concord, MA:
http://welchs.com/company/company_history.html

4. Ibid, ref. 1, p. 55.

5. Knecht, G. Bruce (2006), "The Search for Fresh Beer," The *Wall Street Journal*, January 28.

6. Banwart, George J. (1979), *Basic Food Microbiology,* Westport, Connecticut: AVI Publishing Company, Inc., p. 562.

7. Ibid, ref. 5.

8. Ibid.

9. Ibid.

10. Zanchi, Joseph A. and Alan J. LaBrode (1999), "Combat Ration Logistics From Here to Eternity," *Army Logistician* 31:1, p. 144.

11. Brewin, Bob (2003), "Army to Test Passive RFID Tags on Food Shipments," *Computerworld*, December 1.

12. Ibid, ref. 10.

13. Brock, David L. (2003), "Fresh Food – Dynamic Expiration Dates Using Auto-ID Technology and Analytic Shelf," Cambridge, MA: *MIT Auto-ID Center.*

14. Ibid, ref. 11.

15. Ibid, ref. 13.

16. Ibid, ref. 11.

17. Alien Technololgy, Inc. http://www.alientechnology.com/

18. Ibid, ref. 13.

19. Ibid, ref. 17.

20. Schlesinger, Adam (2005), "Mitigating Container Security Risk Using Real-Time Monitoring with Active Radio Frequency Identification and Sensors," Unpublished Thesis (Master of Engineering), Cambridge, MA: *Massachusetts Institute of Technology.*

21. Emery, Kevin (2005). "Eventing Architecture Considerations: RFID and Sensors in the Supply Chain." Unpublished Thesis (Master of Engineering), Cambridge, MA: *Massachusetts Institute of Technology.*

22. Wasserman, Elizabeth (2006), Keeping Fresh Foods Fresh," *RFID Journal* 3:1, p. 38.

23. Ibid.

24. Ibid.

25. Ibid, p. 33.

26. Ibid, ref. 11.

27. Ibid.

28. Ibid, ref. 22.

Chapter 11 – Retailing: Theft Prevention

1. Earle, Julie (2002), "Inside Track: Retailers turn on the enemy within: Theft," *The Financial Times*, April 3.

2. Two studies attempt to identify the amount of theft as a percentage of total shrinkage (National Retail Security Survey; Efficient Consumer Response report). Both surveys poll loss prevention managers, however, responses are subjective because retailers seldom gather data on the amount of theft. The following table shows responses of both studies.

	US	Europe
Total Shrinkage Costs (Retailers)	$33.2 billion	113.4 billion
Percentage of Sales	1.80%	1.75%
Shoplifting Internal Theft Total Percent Attributed to Theft	31% 46% 77%	37% 24% 61%
The Total Cost of Theft	**$25.6 billion**	**18.17 billion**

The NRSS report also shows that shrinkage, as a percentage of sales, remains stable for the past 10 years. This implies no progress in the reduction in theft.

3. Dow Jones Newswires (2003), "Wal-Mart Pressures Web Site With Bargain Bar Codes," *Dow Jones & Company*, April 18.

4. Ibid, ref. 3.

5. Ibid, ref. 3.

6. Ibid, ref. 2.

7. Albrecht, W. Steve and David I. Searcy (2001),"Top 10 Reasons why Fraud is Increasing in the U.S.," *Strategic Finance* 82:11, p. 58.

8. Alexander, Keith, Tig Gilliam, Kathy Gramling, Chris Grubelic, Herb Kleinberger, Stephen Leng, Dhaval Moogimane, Chris Sheedy (2002), "Applying RFID to Reduce Losses Associated with Shrink," Cambridge, MA: *MIT Auto-ID Center*, p. 13.

9. Ibid, ref. 2.

10. Levin, Amanda (2000), "How RMs Can Curb Cargo Crime," *National Underwriter (Property & Casualty Risk & Benefits Management Edition)* 104:32, p. 3.

11. Wright, Christopher M. (2004), "Somebody Call Security," *APICS – The Performance Advantage*, June.

12. Raman, Ananth, , Nicole DeHoratius, and Zeynep Ton (2001), "The Achilles Heel of Supply Chain Management," *Harvard Business Review* 79:5, p. 25.

13. Zinn, Walter and Peter C. Liu (2001), "Consumer Response to Retail Stockouts," *Journal of Business Logistics* 22:1, p. 59.

14. Lam, Nhat-So (2002), "A Study of the Impact of RFID on Shrinkage Within the Fast Moving Consumer Goods Supply Chain," Unpublished Thesis (Master of Engineering), Cambridge, MA: *Massachusetts Institute of Technology.*

15. Cameron, Mary O. (1964), *The Booster and the Snitch: Department Store Shoplifting,* London: The Free Press of Glencoe, p. 40.

16. Moore, Richard H. (1984), "Shoplifting in Middle America: Patterns and Motivational Correlates," *International Journal of Offenders Therapy and Comparative Criminology* 28:1, pp. 53-64.

17. Post, Richard, ed. (1972), *Combating Crime Against Small Business,* Springfield, Illinois: Charles C. Thomas Publishing, pp. 56-57.

18. Ibid, p. 119.

19. Hayes, Read (1999), "Tailoring Security to Fit the Criminal," *Security Management,* 43:7, pp. 110-116.

20. Sutherland, Edwin Hardin (1937), *The 'Professional' Thief – By a Professional Thief.* Chicago, IL: The University of Chicago Press, pp. 3-40.

21. Meier, Robert F., ed. (1985), *Theoretical Methods in Criminology,* Beverly Hills: Sage Publishing, pp. 151-76.

22. Peacock, Colin, personal correspondence from April 1, 2003.

23. Cox, Anthony D., Dena Cox, Ronald D. Anderson, George P. Moschis (1993), "Social Influences on Adolescent Shoplifting – Theory, Evidence, and Implications for the Retail Industry," *Journal of Retailing* 69:2, pp. 234 – 246.

24. Dawson, Scott (1993), "Consumer Response to Electronic Article Surveillance Alarms," *Journal of Retailing* 69:3, pp. 353 – 363.

24. Ibid, ref. 22.

25. Koh, Robin, Edmund W. Schuster, Nhat-So Lam, and Mark Dinning (2003), "Prediction, Detection and Proof: An Integrated Auto-ID Solution to Retail Theft," Cambridge, MA: *The MIT Auto-ID Center.*

26. Ibid, ref. 22.

27. Baumer, Terry L., and Dennis P. Rosenbaum (1984), *Combating Retail Theft: Programs and Strategies,* Stoneham, Massachusetts: Butterworth Publishers, p. 22.

28. French, Jay T. (1979), *Apprehending and Prosecuting Shoplifters and Dishonest Employees,* New York, NY: National Retail Merchants Association.

29. Beck, A., C. Bilby, P. Chapman, and A. Harrison (2001), *Shrinkage: Introducing a Collaborative Approach to Reducing Stock Loss in the Supply Chain,* Brussels: ECR Europe.

30. Caime, Gabreil and Gabriel Ghone (1996), *S(h)elf Help Guide,* Toronto, Ontario: Trix Publishing.

31. Ibid, ref. 28.

32. Klein, Gerald (1991), "The Hidden Benefits of EAS," *Security Management* 35:6, p. 56.

33. Ibid, ref. 27.

34. Mullen, Fred (1999), "Six Steps to Stopping Internal Theft," *Discount Store News* 38:2, p. 12.

35. Ibid, ref. 19.

36. Hollinger, R., and R. Hayes (2001), "National Retail Security Survey," Gainesville, FL: *University of Florida*.

37. Ibid, ref. 8.

38. Guffey Jr., Hugh J., James R. Harris and J. Ford Laumer Jr. (1979), "Shopper Attitudes Toward Shoplifting and Shoplifting Prevention Devices," *Journal of Retailing* 55:3, p. 75.

39. Ibid, ref. 24.

40. Ibid, ref. 38.

41. Ibid, ref. 22.

42. Ibid, ref. 22.

43. Roberti, Mark (2005), "Retailers Say RFID Will Take Time," *RFID Journal*, January 17.

44. Ibid, ref. 43.

45. Ibid, ref. 43.

46. Ibid, ref. 22.

47. Ibid, ref. 22.

Chapter 12 – Defense: Improving Security and Efficiency

1. Gibbon, Edward (1910), *The Decline and Fall of the Roman Empire – Volume 2*, New York: Alfred A.Knopf, p. 473.
The original text of Gibbon's work was published between 1776 and 1798.

2. Ibid, p. 474.

3. Ibid, p. 493.

4. Ibid, pp. 496 – 511.

5. Ibid, pp. 517 – 518.

6. Morris, Edmund (1979), *The Rise of Theodore Roosevelt*, New York: The Modern Library, p. 656.

7. Ibid, p. 656.

8. Ibid, p. 656.

9. Ibid, p. 657.

10. Andel, Tom (2004), "The Military's War On Invisibility, *Materials Handling Management*, 59:1.

11. Kennedy, Harold (2003), "Electronic Identification Tags Aid Logistics," *National Defense*, 88:597.

12. Saccomano, Ann (2003), No More Weapons of Mass Obstruction," *Journal of Commerce*, May 19.

13. Jackson, William (2003), "Tag Team: Materiel Tracking System Supports Military from Factory to Foxhole," *Government Computing News*, October 13.

14. Ibid, ref. 11.

15. Schmitt, Eric (2004), "The Struggle for Iraq: Official History; Army Study of War Details a 'Morass' of Supply Shortages," *The New York Times*, February 3.

16. Department of Defense (2002), Supply System Inventory Report. *Office of the Under Secretary of Defense for Acquisition, Technology, and Logistics,* http://www.acq.osd.mil/log/logistics_materiel_readiness/organizations/sci/ assetts/executive_info/ssir_new/2002/2002ssir.pdf

17. Shelton, Henry H. (2000), "JV 2020," Washington, D.C.: *U.S. Government Printing Office.*

18. Kaplan, Robert, D. (2004), "Indian Country," *The Wall Street Journal*, September 21.

19. Erwin, Sandra I. (2004), "Changes on the Way for Army Logistics Ops," *National Defense*, April 29.

20. Lai, Elaine M. (2003), An Analysis of the Department of Defense Supply Chain: Potential Applications of the RFID Center Technology to Improve Effectiveness, Unpublished Thesis (Bachlers of Science), Cambridge: MA: *Massachusetts Institute of Technology.*

21. RFID Technology: Transportation and Logistics Adoption Forum (2002), Center for Transportation and Logistics, Cambridge, MA: *Massachusetts Institute of Technology.*

22. Ibid, ref 17.

23. Engels, Daniel M., Robin Koh, Elaine M. Lai, and Edmund W. Schuster (2004), "Improving Visibility in the DOD Supply Chain," *Army Logistician* 36:3.

24. Ibid, ref. 20.

25. Coyle, John, J., Edward J. Bardi, and C. John Langley, Jr. (1992), *The Management of Business Logistics*, New York: West Publishing Company, pp. 329 – 339.

26. Alexander, Saraj M. (1985), "Discovering and Correcting Problems in a Navel Stock Control System," *Interfaces* 15:4.

27. Fricker, Ronald D., Jr., and Marc Robbins (2000), "Retooling for the Logistics Revolution: Designing Marine Corps Inventories to Support the Warfighter," Technical report, MR-1096-USMC, Santa Monica, CA: *The RAND Corporation.*

28. Coyle, R. G. and Paul A. Gardiner (1991), "A System Dynamics Model of Submarine Operations and Maintenance Schedules," *Journal of the Operational Research Society* 42:6.

29. Ibid, ref. 20.

30. Ibid, ref. 20.

31. Ibid, ref. 20.

32. Ibid, ref. 20.

33. Harry Kirejczyk (2003), Private interview, operations research analyst, modeling and analysis team, *U.S. Army Natick Soldier Center.*

34. Ibid, ref. 12.

35. Ibid, ref. 20.

36. Cottrill, Ken (2003), "Savi Wins DOD Contract," *Traffic World*, February 17.

37. Carlton, Jim (2003), "Savi Technology is Awarded $90 Million Pentagon Contract," *The Wall Street Journal*, February 6.

38. Hughes, David (1996), "Electronic Tags Track Cargo Bound for Bosnia," *Aviation Week & Space Technology*, 144:1 January 1.

39. Bacheldor, Beth (2003), "Defense Department to Boost RFID Adoption," *InformationWeek*, October 13, 2003.

40. Whiting, Rick (2004), "Alliance To Use Technology to Manage Supplies from Europe to Afghanistan," *InformationWeek*, March 29.

41. Brewin, Bob (2005), "For DOD, "One in 1,048,000 is Not Unique Enough," *Federal Computer Week*, March 25.

42. Ibid

43. Ibid, ref. 21.

44. Roberti, Mark (2004), "DOD Clarifies UID vs. EPC Issue, *RFID Journal*, April. 8.

45. Johnson, George (2005), Who do you Trust More: G.I. Joe or A.I. Joe, *The New York Times*, February 20.

Chapter 13 – The Role of Data in Enterprise Resource Planning

1. Joshi, Yogesh V. (2000), "Information Visibility and its Effect On Supply Chain Dynamics," Unpublished Thesis (Master of Science), Cambridge, MA: *Massachusetts Institute of Technology.*

2. Wise, Richard and Peter Baumgartner (1999), "Go Downstream," *Harvard Business Review,* September-October.

3. McFarlane, Duncan and Yossi Sheffi (2003), "The Impact of automatic Identification on Supply Chain Operations," *International Journal of Logistics Management* 14:1, p. 2.

4. Maher, Kris (2005), "Global Goods Jugglers," *The Wall Street Journal,* July, 5.

5. Bornhövd, Christof, Tao Lin, Stephan Haller and Joachim Schaper (2004), "Integrating Automatic Data Acquisition with Business Processes: Experiences with SAP's Auto-ID Infrastructure," *Proceedings of the 30th VLDB Conference*: Toronto: Canada.

6. Miles, Robert H. (1980), *Macro Organizational Behavior*, Glenview, Illinois: Scott, Foresman and Company, p. 199.

7. Allen, Stuart J. and Edmund W. Schuster (2004), "Controlling the Risk for an Agricultural Harvest," *Manufacturing & Service Operations Management* 6:3, pp 1 – 12.

8. Schuster, Edmund, W., David L. Brock, Stuart J. Allen, Pinaki Kar, and Mark Dinning (2005), "Enabling ERP through Auto-ID Technology." In *Strategic ERP Extension and Use,* edited by E. Bendoly and F.R. Jacobs. Palo Alto: Stanford University Press.

9. Bostwick, Peter (2004), "Method and System for Creating ,Sustaining and Using a Transactional Bill of Materials (TBOM ™)," Washington, D.C.: *U.S. Patent Office.*

10. Engels, Daniel W., Robin Koh, Elaine M. Lai, and Edmund W. Schuster (2004), "Improving Visibility in the DOD Supply Chain," *Army Logistician* 36:3, pp. 20 – 23.

11. Kar, Pinaki, Ming Li, and Edmund W. Schuster (2003), "A Case Study of Computer Service Parts Inventory Management," Service Parts Resource Guide (1st Edition), Falls Church, VA: *American Production and Inventory Control Society.*

12. Nahmias, Steven (1993), *Production and Operations Analysis*, Boston: Irwin.

13. McFarlane, Duncan (2002), "Auto-ID Based Control," Cambridge, MA: *MIT Auto-ID Center.*

14. McFarlane, Duncan (2003), "The Impact of Product Identity on Industrial Control Part 1: 'See More, Do More…,'" Cambridge, MA: *MIT Auto-ID Center.*

15. Schuster, Edmund W. and Byron J. Finch (1990), "A Deterministic Spreadsheet Simulation Model for Production Scheduling in a Lumpy Demand Environment," *Production and Inventory Management Journal* 31:1, pp.39-42.

16. Allen, Stuart .J. and Edmund W. Schuster (1994), "Practical Production Scheduling With Capacity Constraints and Dynamic Demand: Family Planning and Disaggregation," *Production and Inventory Management Journal* 35:4, pp. 15-21.

17. Allen, Stuart J., Jack Martin and Edmund W. Schuster (1997), "A Simple Method for the Multi-Item, Single Level, Capacitated Scheduling Problem with Set-Up Times and Costs," *Production and Inventory Management Journal* 38:4, p. 39-47.

18. D'Itri, Michael P., Stuart J. Allen and Edmund W. Schuster (1999), "Capacitated Scheduling of Multiple Products on a Single Processor with Sequence Dependencies," *Production and Inventory Management Journal*, 40:5, pp. 27 – 33.

19. Billington, P.J., McClain J.O., and Thomas, L.J. (1983), "Mathematical Programming Approaches To Capacity-Constrained MRP Systems: Review, Formulation and Problem Reduction," *Management Science* 29:10, pp. 1126.

20. Taylor, Sam G. and Steve F. Bolander (1994), *Process Flow Scheduling: A Scheduling Systems Framework for Flow Manufacturing*, Falls Church, VA: American Production and Inventory Control Society, p. 1.

21. Schuster, Edmund W., Chatchai Unahabhokha, Stuart J. Allen (2003), "Master Production Schedule Stability under Conditions of Finite Capacity," Working Paper, Cambridge, MA: *MIT Laboratory for Manufacturing and Productivity - Data Center Program*.

22. Schuster, Edmund W. and Stuart J. Allen (1998), "Raw Material Management at Welch's," *Interfaces* 28:5, p. 13-24.

23. Schuster, Edmund W., Stuart J. Allen and Michael P. D'Itri, M.P (2000), "Capacitated Materials Requirements Planning and its Application in the Process Industries," *Journal of Business Logistics*, 21:1, pp. 169 – 189.

24. Leachman, R.C., Benson, R.F., Liu, C. and Raar, D.J. (1996), "IMPReSS: An Automated Production-Planning and Delivery-Quotation System at Harris Corporation-Semiconductor Sector," *Interfaces* 26:1, p. 6-37.

25. Chappell, Gavin, Lyle Ginsburg, Paul Schmidt, Jeff Smith, and Joseph Tobolski (2003), "Auto-ID on the Line: The Value of Auto-ID Technology in Manufacturing," Cambridge, MA: *MIT Auto-ID Center*.

26. Ibid, ref. 5.

Chapter 14 – Building a Business Case for the EPCglobal Network

1. McGee, Ken (2004), "Give Me That Real-Time Information," *Harvard Business Review* 82:4, p. 26.

2. Gardner, W. David (2004), "Analyst: Wal-Mart's RFID Suppliers Are Resisting," *TechWeb News*, Dec. 21.

3. Malykhina, Elena (2004), "RFID's Payback Stretches Beyond 2 Years, Analyst," *InformationWeek*, October 29.

4. Dinning, Mark and Edmund W. Schuster (2004), "Getting on Board: Building a Business Case for RFID at Dell," *APICS – The Performance Advantage*, October.

5. Hadow, Robert (2005), The Math Behind RFID in Logistics," *RFID Journal*, January 31.

6. Bacheldor, Beth (2004), "RFID Investment Will Grow, But ROI Will Take Time," *InformationWeek*, May 17.

7. Laubacher, Robert, S. P. Kothari, Thomas W. Malone, Brian Subirana (2005), "What is RFID Worth to your Company? Measuring Performance at the Activity Level," Cambridge, MA: *MIT Sloan School of Management*.

8. Lee, Hau L. (2000), "Creating Value Through Supply Chain Integration," *Supply Chain Management Review*, September 1.

9. Hoppe, Richard (2001), "Outlining a Future of Supply Chain Management – Coordinated Supply Networks," Unpublished Thesis (Master of Science), Cambridge, MA: *Massachusetts Institute of Technology*.

10. Sabath, Robert, Chad W. Autry, and Patricia J. Daugherty (2001), "Automatic Replenishment Programs: The Impact of Organizational Structures," *Journal of Business Logistics* 22:1.

11. Ibid, ref. 9.

12. Simchi-Levi, David, Philip Kaminsky, and Edith Simchi-Levi (2000), *Designing and Managing the Supply Chain: Concepts, Strategies and Case Studies*, Boston: Irwin McGraw-Hill, pp. 92 – 93.

13. Rice, James B. and Richard M. Hoppe (2001), "Supply Chain vs Supply Chain: The Hype & the Reality," *Supply Chain Management Review*, September/October.

14. Ibid, ref. 7.

15. Ibid, ref. 7.

16. Stewart, Thomas A. and Louise O'Brien (2005), "Execution Without Excuses: The HBR Interview, Michael Dell and Kevin Rollins," *Harvard Business Review*, March.

17. Ibid.

18. Magretta, Joan (1998), "The Power of Virtual Integration: An Interview with Dell Computer's Michael Dell," *Harvard Business Review*, March-April.

19. Ibid.

20. Ibid, ref. 16.

21. Ibid.

22. Ibid.

23. Sennechael, Denis and Iain MaClean (2001), "Car Sequencing Challenge at a Nissan Plant Calls for Complex Scheduling," *APICS – The Performance Advantage*, March.

24. Ibid ref. 16.

25. Liberatone, Matthew J. and Tan Miller (1998), A Framework for Integrating Activity-based Costing and The Balanced Scorecard into The Logistics Strategy Development and Monitoring Process," *Journal of Business Logistics* 19:2.

26. Evans, Bob (2005), Business Technology: Implementing RFID is a Risk Worth Taking, *InformationWeek*, June 13.

27. Chabrow, Eric (2005), "VoIP And Security Beat Out RFID In Battle For Business Dollars," *InformationWeek*, March 21.

28. Roberti, Mark (2004), "GMA:Business Case for EPC Mixed," *RFID Journal*, Nov. 22.

Chapter 15 – Enhancing Revenue Using the EPC

1. Attributed to Everett N. Baldwin, CEO of Welch Foods, Inc., 1982 – 1995.

2. Henry, David and Frederick F. Jespersen (2002), "Mergers, Why Most Big Deals do not Pay Off," *BusinessWeek*, October 14.

3. Bucklin, Randolph E. and Sunil Gupta (1999), "Commercial Use of UPC Scanner Data: Industry and Academic Perspectives," *Marketing Science* 18:3.

4. Hanssens, Dominique M., Leonard J. Parsons and Randall L. Schultz (2001), *Market Response Models, 2nd Edition*, Boston, MA: Kluwer Academic Publishers.

5. Brody, Edward, I. (2001), Marketing Engineering at BBDO," *Interfaces* 31:3.

6. Schoenberger, Chana R. (2002), "The Internet of Things," *Forbes*, March 18.

7. Hunter, Paul and Mark Hinds (2005), "Making Sense of Data: Customers at the Center of your Business," Engineering Marketing Science, Cambridge, MA: *MIT Laboratory for Manufacturing and Productivity - Data Center Program*.

8. Unauthored (2005), "CVS Halts Access Over Web to Data of Loyalty Cards," *The Associated Press*, June 22.

9. Montgomery, Alan L. (2001), "Applying Quantitative Marketing Techniques to the Internet," *Interfaces* 31:2.

10. Hanssens, Dominique M. (2005), "Market Response Models and Demand Creation," Engineering Marketing Science, Cambridge, MA: *MIT Laboratory for Manufacturing and Productivity - Data Center Program*.

11. Vranica, Suzanne (2005), "*Advertising* Anywhere, Anytime," *The Wall Street Journal*, November 21.

12. Merrick, Amy (2005), "Gap's Marketing Head Steps Down," *The Wall Street Journal*, December 2.

13. Ellinson, Sarah (2005), "Anheuser Will Raise Spending On Cable, Internet," *The Wall Street Journal*, November 30.

14. Angwin, Julia and Kevin J. Delaney (2005), Top Web Sites Build Up Ad Backlog, Raise Rates, *The Wall Street Journal*, November 16.

15. Ibid.

16. Boslet, Mark (2005), "Advertisers Welcome Possible Microsoft-AOL Deal," *Dow Jones NewsWires*, December 9.

17. Ibid.

18. Patrick, Aaron O. (2005), "Yahoo to Track Impact of Internet Ads," *The Wall Street Journal*, December 16.

19. Ibid.

20. Bosman, Julie (2005), "Would You Like Some Fries With That Download?" *The New York Times*, December 12.

21. Ibid.

22. Bawa, Kapil, Janet T. Landwehr, and Ardadhna Krishna (1989), "Consmer Response to Reatailers' Marketing Environments: An analysis of Coffee Purchase Data," *Journal of Retailing* 65:4, pp. 492.

23. Nelson, Emily and Sarah Ellison (2005), "In a Shift, Marketers Beef Up Ad Spending Inside Stores," *The Wall Street Journal*, September 21.

24. Sheng, Ellen (2005), "Signs of the Times: Digital Ads Allow Marketers to Target Pitches," *Dow Jones News Wires*, October 26.

25. Ibid.

26. Ibid.

27. Collins, Jonathan (2004), Checkpoint Backs Goliath," *RFID Journal*, March 2.

28. Fowler, Geoffry (2005), "In Japan, Billboards Take Code-Crazy Ads to New Heights," *The Wall Street Journal*, October 10.

29. Ball, Deborah (2004), "Consumer-Goods Firms Duel for Shelf Space," *The Wall Street Journal*, October 22, pp B2.

30. Roberts, William A. (2003), "2001 New Products Conference: Take a Chance," *Prepared Foods*, November 21.

31. Desiraju, Romarao (2001), "New Product Introductions, Slotting Allowances, and Retailer Discretion," *Journal of Retailing* 77, pp. 335 – 358.

32. Lariviere, Martin A. and V. Padmanabhan (1997), "Slotting Allowances and New Product Introductions," *Marketing Science* 16:2.

33. Brock, David L., Edmund W. Schuster, Stuart J. Allen and Pinaki Kar (2005), "An Introduction to Semantic Modeling for Logistical Systems," *Journal of Business Logistics* 26:2, pp. 97 – 117.

34. Allaway, Arthur W., David Berkowitz and Giles D'Souza (2003), "Spatial Diffusion of a New Loyalty Program Through a Retail Market," *Journal of Retailing*, Vol. 79, pp 137 – 151.

35. Bronnenberg, Bart J. and Carl F. Mela (2004), "Market Roll-Out and Retailer Adoption for New Brands," *Marketing Science*, Vol. 23, No. 4, pp. 500 – 518.

36. Skarra, Leslie (2004), "Rollout Roulette," *Prepared Foods*, May 9.

37. Urban, Glen L. (1993), *Design and Marketing of New Products*, Englewood Cliffs, N.J.: Prentice Hall.

38. Belloni, Alexandre, Robert Freund, Matt Selove, and Duncan Simester (2005), "Optimizing Product Line Design," Engineering Marketing Science, Cambridge, MA, *MIT Laboratory for Manufacturing and Productivity - Data Center Program.*

39. Schuster, Edmund W. (2004), *Private Industry Correspondence Relating to Market Research Data.*

40. Fotedar, Shivi, Mario Gerla, Paola Crocetti, and Luigi Fratta (1995), "ATM Virtual Private Networks," *Communications of the ACM* 38:2, pp. 101 – 109.

41. Dichter, Ernest (1966), "How Word of Mouth Advertising Works," *Harvard Business Review*, Nov. – Dec., p. 147.

42. Bernard, Tara Siegel (2005), "Small Firms Turn to Marketing Buzz Agents," *The Wall Street Journal*, December 27.

43. Ryan, Bryce and Gross, Neal C. (1943), "The Diffusion of Hybrid Seed Corn in Two Iowa Communities," *Rural Sociology* 8, pp. 15-24.

44. Bass, Frank M. (1969), "A new product growth model for consumer durables," *Management Science* 15:1, pp. 215–227.

45. Ibid, ref. 9.

46. Ibid, ref, 9.

47. Golder, Peter N. and Gerard J. Tellis (1997), "Will It Ever Fly? Modeling The Growth of New Consumer Durables," *Marketing Science*, 16, 3, 256-270.

48. Tellis, Gerard J., Stefan Stremersch and Eden Yin (2003), "The International Takeoff of New Products: Economics, Culture and Country Innovativeness,"*Marketing Science*, 22:2, pp. 161-187.

49. Unauthored (2005), "Immelt Sees 60% Of GE Growth Coming From Developing World," *Dow Jones Newswires*, February 4.

50. Fujita, Masahisa, Paul Krugman, and Anthony J. Venables (2000), *The Spatial Economy*, Cambridge, MA: The MIT Press.

51. Whyte, William H., Jr. (1954), "The Web Word of Mouth," *Fortune*, November, pp. 140.

52. Ibid, ref. 34.

53. Garber, Tal, Jacob Goldenberg, Barak Libai, and Eitan Muller (2004), "From Density to Destiny: Using Spatial Dimension of Sales Data for Early Prediction of New Product Success," *Marketing Science*, Vol. 23, No. 3, pp. 419-428.

54. Zinn, Walter and Peter C. Liu (2001), "Consumer Response to Retail Stockouts," *Journal of Business Logistics*, Vol. 22, No. 1, pp. 49-71.

55. Ibid, ref. 34.

56. Curry, David J. (1989), "Single-Source Systems: Retail Management Present and Future," *Journal of Retailing* 65:1, pp. 1 – 20.

57. See Homescan by A.C. Neilsen

58. Ibid.

59. Shugan, Steven M. (2002), "In Search of Data: an Editorial," *Marketing Science* 21:4, p.371.

60. Hays, Constance L. (2004), "What Wal-Mart Knows About Customers' Habits," *The New York Times*, November 14.

61. Kang, Stephanie (2005), "Retail Spending Rose 8.7% in Holiday '05," *The Wall Street Journal*, December 27.

62. Hudson, Kris and Mylene Mangalind (2005), "U.S. Retailers Ring up Strong Weekend Sales," *The Wall Street Journal*, November 27.

63. Ibid, ref. 34.

64. Ibid, ref. 8.

65. Kontzer, Tony (2005), RFID – Future Consumer Data Battleground," *Information-Week*, Aug. 18.

66. Chandrasekhar, Chaitra and Edmund W. Schuster (2005), "Design of User Interfaces for Computing Systems," *Cutter Consortium Executive Update*.

67. GoodChild, Michael F. (1991), "Geographic Information Systems," *Journal of Retailing* 67:1, pp. 3 – 14.

68. Earth Resource Surveys, Inc., http://www.ersi.ca/

69. Eisenberg, Anne (2005), "What's Next; A 3-D View of the City, Block by Block," *The New York Times*, February 17.

70. Auchard, Eric (2004), "Yahoo Maps Offer Live US Traffic Conditions," *Reuters*, December 16.

71. Gomes, Lee (2005), "Google Pioneers Use Old Microsoft Tools in New Web Programs," *The Wall Street Journal*, March 14.

72. Unauthored (2005), "Google Combines Online Maps With Local Search Features," *Dow Jones Newswires*, October 6.

73. Saranow, Jennifer and Mylene Mangalindan (2005), "Getting an Oil Change Off eBay," *The Wall Street Journal*, November 17.

74. Delaney, Kevin J. (2005), "Internet Ads Click with Firms; Some Shift Budgets," *The Wall Street Journal*, May 3.

75. Dash, Eric (2005), "A Web Site Maps Home Searches," *The New York Times*, May 1.

76. Waters, Richard (2005), "Plugging Together Software May Soon be Painless," *Financial Times*, May 3.

77. Rust, Roland T. and Julia A.N. Crown (1986), "Estimation of Marketing Area Densities," *Journal of Retailing* 62:4, pp. 410 – 430.

78. Delaney, Kevin J. and Jesse Drucker (2005), "Google Proposes San Francisco Wi-Fi," *The Wall Street Journal*, September 30.

Chapter 16 – Outlook: Navigating the Sea of Data

1. Gupta, Sunil. (2004), "Empowering a Consumer Driven Demand Chain with Global Data Synchronization," *Smart World 2004 -- Semantic Modeling*: Cambridge, MA, December 8.

2. Park, Andrew (2004), "Can EMC Find Growth Beyond Hardware?" *BusinessWeek*, November 1.

3. Lyons, Daniel (2004), "Too Much Data," *Forbes*, December 13.

4. Ibid.

5. Ibid, ref. 2.

6. Whiting, Rick and Tony Kontzer (2004), "Data Avalanche," *InformationWeek*, February 16.

7. Ferguson, Glover, Sanjay Mathur and Baiju Shah (2005), "Evolving From Information to Insight," *Sloan Management Review* 46:2, p. 52.

8. Brock, David L., Edmund W. Schuster, Stuart J. Allen, and Pinaki Kar (2005), "An Introduction to Semantic Modeling for Logistical Systems," *Journal of Business Logistics* 26:2, pp. 97 – 117.

9. Attributed to Professor Gregor M. Reinhard of Gannon University, GH-501 Pubic Policy Process, taught in 1984.

10. Forrester, Jay W. (1961), *Industrial Dynamics*, Waltham, MA: Pegasus Communications.

11. Willemain, Thomas R. (1994), "Insights on Modeling from a Dozen Experts," *Operations Research* 42:2, pp. 213-222.

12. Willemain, Thomas R. (1995), "Model Formulation: What Experts Think about and When," *Operations Research* 43:6, pp. 916-932.

13. Geoffrion, Arthur M. (1987), "An Introduction to Structured Modeling," *Management Science* 33:5, pp. 547-588.

14. Brock, David L. (2003), "Beyond the EPC – Making Sense of the Data – Proposals for Engineering the Next Generation Intelligent Data Network," *The MIT Auto-ID Center*: Cambridge, MA.

15. Brock, D.L., "The Intelligent Data Network Proposal for Engineering the Next Generation of Distributed Data Modeling, Analysis and Prediction," Cambridge, MA: *MIT Laboratory for Manufacturing and Productivity - Data Center Program*.

16. Fine, Charles H. (1998), *Clockspeed*, Reading, MA: Perseus Books.

17. Lederer, Richard, *The Miracle of Language*, New York: Simon & Schuster, 1991, p. 16.

18. Fensel, Dieter, James Hendler, Henry Lieberman and Wolfgang Wahlster (2003), *Spinning the Semantic Web*, Cambridge, MA: MIT Press, p. 377.

19. Ibid, p. 7.

20. Schuster, Edmund W. and David L. Brock (2004), "Creating an Intelligent Infrastructure for ERP Systems; The Role of RFID Technology," Falls Church, VA: *American Production and Inventory Control Society*.

21. Deacon, Terrence, W. (1997), *The Symbolic Species*, New York: W.W. Norton, p. 48.

22. Howard, Philip (2005), "A Way with Words," *The Wall Street Journal*, August 23.

23. Attributed to David L. Brock of MIT.

24. The M dictionary allows 'compound' words such as motor_vehicle.1, though a legitimate question would be whether this should be a phrase using two words from the dictionary (i.e. motor.1_vehicle.1). In general there is no rule, but as a guideline compound words with unique meaning, such as "public relations" or "accounts receivable," or common pairings, such as "telephone number" should be compound words in the dictionary (i.e. public_relations.1, accounts_receivable.1, and telephone_number.1).
Conversely, phrases whose meanings can be inferred from their parts, such as 'red ball' or 'patient name', should be formed as phrases (i.e. red.1_ball.1 and patient.1_name.1). This allows complex communication without the resulting combinatorial explosion that would occur from storing all word sequences.

25. WordNet®, Princeton University, http://wordnet.princeton.edu/.

26. Ibid.

27. Wiki, http://en.wikipedia.org/wiki/Wiki.

28. Unauthored (2005), "Science Journal Says Wikipedia is Fairly Accurrate," *Associated Press*, December 15.

29. Wessel, Rhea (2005), "Office Technology Opening Up," *The Wall Street Journal*, October 24.

30. Wikipedia, http://en.wikipedia.org

31. Grosof, Benjamin N. (2005), "Lecture on Semantic Web and Semantic Web Services," Cambridge, MA: *Massachusetts Institute of Technology*, April 5.

32. Ibid, ref. 17, p. 102

33. Gibbon, Edward (1910), *The Decline and Fall of the Roman Empire – Volume 6*, New York: Alfred A.Knopf, p. 493.

Glossary[f]

Adaptive Control

1) The ability of a control system to change its own parameters in response to a measured change in operating conditions. 2) Machine control units in which feeds and/or speeds are not fixed. The control unit, working from feedback sensors, is able to optimize favorable situations by automatically increasing or decreasing the machining parameters. This process ensures optimum tool life or surface finish and/or machining costs or production rates.†

Advance Ship Notice (ASN)

An EDI notification of shipment of product. †

Advanced Planning and Scheduling (APS)

Techniques that deal with analysis and planning of logistics and manufacturing over the short, intermediate, and long-term time periods. APS describes any computer program that uses advanced mathematical algorithms or logic to perform optimization or simulation on finite capacity scheduling, sourcing, capital planning, resource planning, forecasting, demand management, and others. These techniques simultaneously consider a range of constraints and business rules to provide real-time planning and scheduling, decision support, available-to-promise, and capable-to-promise capabilities. APS often generates and evaluates multiple scenarios. Management then selects one scenario to use as the "official plan." The five main components of APS systems are demand planning, production planning, production scheduling, distribution planning, and transportation planning. †

Agile Reader

Generic term for a reader that can read different types of RFID tags, such as those made by different manufacturers, or those that operate at different frequencies.*

---

f †From the APICS Dictionary
  *From the Auto-ID Center Web Site

Algorithm

A prescribed set of well-defined rules or processes for solving a problem in a finite number of steps, e.g., the full statement of the arithmetic procedure for calculating the reorder point. †

American National Standards Institute (ANSI)

The parent organization of the interindustry electronic interchange of the business transaction standard. This group is the clearinghouse on U.S. electronic data interchange standards. †

Amplitude

The maximum height of a radio wave. *

Analog Data

Information that is represented by continuously changing physical quantity, such as length or height of an electromagnetic wave (see below). *

Analog

As applied to an electrical or computer system, the capability of representing data in continuously varying physical phenomena (as in a voltmeter) and converting them into numbers. †

ANSI

Acronym for American National Standards Institute. †

Antenna

A device for sending or receiving electromagnetic waves. *

Anti-collision

A technique used to prevent several tags in the field of a single reader, or readers with overlapping fields, from interfering with one another. Anti-collision algorithms typically work by ensuring that the tags or readers don't transmit at the same time. *

APICS

The Educational Society for Resource Management : Founded in 1957 as the American Production and Inventory Control Society, APICS is a not-for-profit educational organization consisting of 70,000 members in the production/operations, materials, and integrated resource management areas. †

Application Package

A computer program or set of programs designed for a specific application; e.g., inventory control, MRP. †

Application System

A set of programs of specific instructions for processing activities needed to compute specific tasks for computer users, as opposed to operating systems that control the computer's internal operations. Examples are payroll, spreadsheets, and word processing programs. †

Applications Software

Programs created for a particular business purpose such as payroll or inventory control. †

ASN

Abbreviation for advance ship notice. †

Automated Data Capture System

Any device such as a bar-code reader or optical character reader that mechanizes the entry of information into an information system. †

Automatic Data Capture (ADC)

Methods of collecting data and entering it directly into computer systems without human involvement (see also automatic identification and data collection).*

Automatic Identification and Data Collection (AIDC)

A broad term that covers methods of entering data directly into a computer system without using a keyboard. These include barcode scanning, radio frequency identification, voice recognition and other technologies. *

Automated Information System (AIS)

Computer hardware and software configured to automate calculating, computing, sequencing, storing, retrieving, displaying, communicating, or otherwise manipulating data and textual material to provide information. †

Automation

The substitution of machine work for human physical and mental work, or the use of machines for work not otherwise able to be accomplished, entailing a less continuous interaction with humans than previous equipment used for similar tasks. †

Backflush

A method of inventory bookkeeping where the book (computer) inventory of components is automatically reduced by the computer after completion of activity on the component's upper-level parent item based on what should have been used as specified on the bill of material and allocation records. This approach has the disadvantage of a built-in differential between the book record and what is physically in stock. Syn: explode-to-deduct, post-deduct inventory transaction processing. See: pre-deduct inventory transaction processing. †

Bandwidth

In telecommunications, a measurement of how much data can be moved along a communications channel per unit of time, usually measured in bits per second. †

Bar Code

A standard adopted to make it possible for machines to automatically identify labeled objects. The barcode was adopted because the bars were easier for machines to read than characters that humans could read. The main drawbacks of the bar code system in common use are that it can't distinguish one can of soup from another and scanners have to have line of sight to read the label. *

Bar Coding

A method of encoding data using bar code for fast and accurate readability. †

Baud

The number of bits transmitted per second. †

Bit

The smallest unit of digital information – a single one or zero. A 96-bit EPC is a string of 96 ones and zeros. *

BOM

Abbreviation for bill of material. †

BPR

Abbreviation for business process reengineering. †

Broadband

A coaxial cable offering several channels for text, voice, and/or video transmission. †

Browser

Software used on the Web to retrieve and display documents on-screen, connect to other sites using hypertext links, display images, and play audio files. †

Bullwhip Effect

An extreme change in the supply position upstream in a supply chain generated by a small change in demand downstream in the supply chain. Inventory can quickly move from being backordered to being excess. This is caused by the serial nature of communicating orders up the chain with the inherent transportation delays of moving product down the chain. The bullwhip effect can be eliminated by synchronizing the supply chain. †

Business Process Reengineering (BPR)

A procedure that involves the fundamental rethinking and radical redesign of business processes to achieve dramatic organizational improvements in such critical measures of performance as cost, quality, service, and speed. Any BPR activity is distinguished by its emphasis on (1) process rather than functions and products and (2) the customers for the process. Syn: reengineering. †

Byte

A string of 8 bits used to represent a single character in a computer code. †

Cache

A high-speed device used within a computer to store frequently retrieved data. †

Capacity

1) The capability of a system to perform its expected function. 2) The capability of a worker, machine, work center, plant, or organization to produce output per time period. Capacity required represents the system capability needed to make a given product mix (assuming technology, product specification, etc.). As a planning function, both capacity available and capacity required can be measured in the short term (capacity requirements plan), intermediate term (rough-cut capacity plan), and long term (resource requirements plan). Capacity control is the execution through the I/O control report of the short-term plan. Capacity can be classified as budgeted, dedicated, demonstrated, productive, protective, rated, safety, standing, or theoretical. See: capacity available, capacity required. 3) Required mental ability to enter into a contract. †

Capacity Management

The function of establishing, measuring, monitoring, and adjusting limits or levels of capacity in order to execute all manufacturing schedules; i.e., the production plan, master production schedule, material requirements plan, and dispatch list. Capacity management is executed at four levels: resource requirements planning, rough-cut capacity planning, capacity requirements planning, and input/output control. †

Cargo

A product shipped in an aircraft, railroad car, ship, barge, or truck. †

Carrying Cost

The cost of holding inventory, usually defined as a percentage of the dollar value of inventory per unit of time (generally one year). Carrying cost depends mainly on the cost of capital invested as well as such costs of maintaining the inventory as taxes and insurance, obsolescence, spoilage, and space occupied. Such costs vary from 10% to 35% annually, depending on type of industry. Carrying cost is ultimately a policy variable reflecting the opportunity cost of alternative uses for funds invested in inventory. Syn: holding costs. †

Category Management

In marketing, an organizational structure giving managers responsibility for planning and implementing marketing systems for certain product lines. †

Cellular Manufacturing

A manufacturing process that produces families of parts within a single line or cell of machines controlled by operators who work only within the line or cell. †

Central Processing Unit (CPU)

The electronic processing unit of a computer, where mathematical calculations are -performed. †

Channels of Distribution

Any series of firms or individuals that participates in the flow of goods and services from the raw material supplier and producer to the final user or consumer. See: distribution channel. †

Check Digit

A digit added to each number in a coding system that allows for detection of errors in the recording of the code numbers. Through the use of the check digit and a predetermined mathematical formula, recording errors such as digit reversal or omission can be discovered. †

Chip

See microchip. *

Closed-Loop Feedback System

A planning and control system that monitors system progress toward the plan and has an internal control and replanning capability. †

Closed-Loop MRP

A system built around material requirements planning that includes the additional planning processes of production planning (sales and operations planning), master production scheduling, and capacity requirements planning. Once this planning phase is complete and the plans have been accepted as realistic and attainable, the execution processes come into play. These processes include the manufacturing control processes of input-output (capacity) measurement, detailed scheduling and dispatching, as well as anticipated delay reports from both the plant and suppliers, supplier scheduling, and so on. The term closed loop implies not only that each of these processes is included in the overall system, but also that feedback is provided by the execution processes so that the planning can be kept valid at all times. †

Collaborative Planning

Syn: collaborative planning, forecasting, and replenishment. †

Collaborative Planning, Forecasting, and Replenishment (CPFR)

1) A collaboration process whereby supply chain trading partners can jointly plan key supply chain activities from production and delivery of raw materials to production and delivery of final products to end customers. Collaboration encompasses business planning, sales forecasting, and all operations required to replenish raw materials and finished goods. 2) A process philosophy for facilitating collaborative communications. CPFR is considered a standard, endorsed by the Voluntary Interindustry Commerce Standards. Syn: collaborative planning. †

Collision

Radio signals interfering with one another. Signals from tags or readers can collide (see below). *

Competitive Advantage

An edge, e.g., a process, patent, management philosophy, or distribution system, that a seller has that enables the seller to control a larger market share or profit than the seller would otherwise have. Syn: competitive edge. †

Competitive Analysis

An analysis of a competitor that includes its strategies, capabilities, prices, and costs. †

Competitive Intelligence

The information required to conduct a competitive analysis.

Connectivity

The ability to communicate effectively with supply chain partners to facilitate interorganization synchronization. †

Constraint

Any element or factor that prevents a system from achieving a higher level of performance with respect to its goal. Constraints can be physical, such as a machine center or lack of material, but they can also be managerial, such as a policy or procedure. †

Container

A large box in which commodities to be shipped are placed. †

Containerization

A shipment method in which commodities are placed in containers, and after initial loading, the commodities per se are not rehandled in shipment until they are unloaded at the destination. †

Continuous Replenishment

A process by which a supplier is notified daily of actual sales or warehouse shipments and commits to replenishing these sales (by size, color, and so on) without stockouts and without receiving replenishment orders. The result is a lowering of associated costs and an improvement in inventory turnover. See: vendor-managed inventory. †

Control System

A system that has as its primary function the collection and analysis of feedback from a given set of functions for the purpose of controlling the functions. Control may be implemented by monitoring or systematically modifying parameters or policies used in those functions, or by preparing control reports that initiate useful action with respect to significant deviations and exceptions. †

Coupling

The transfer of energy from one electronic circuit to another. Inductive and capacitive coupling are two methods used to transfer energy (and also data) between a reader and a tag. *

CRM

Abbreviation for customer relationship management and customer relations management. †

Cross-Docking

The concept of packing products on the incoming shipments so they can be easily sorted at intermediate warehouses or for outgoing shipments based on final destination. The items are carried from the incoming vehicle docking point to the outgoing vehicle docking point without being stored in inventory at the warehouse. Cross-docking reduces inventory investment and storage space requirements. Syn: direct loading. †

Customer Relationship Management (CRM)

A marketing philosophy based on putting the customer first. The collection and analysis of information designed for sales and marketing decision support (as contrasted to enterprise resources planning information) to understand and support existing and potential customer needs. It includes account management, catalog and order entry, payment processing, credits and adjustments, and other functions. Syn: customer relations management. †

Customer Service Ratio

1) A measure of delivery performance of finished goods, usually expressed as a percentage. In a make-to-stock company, this percentage usually represents the number of items or dollars (on one or more customer orders) that were shipped on schedule for a specific time period, compared with the total that were supposed to be shipped in that time period. Syn: customer service level, fill rate, order-fill ratio, percent of fill. Ant: stockout percentage. 2) In a make-to-order company, it is usually some comparison of the number of jobs or dollars shipped in a given time period (e.g., a week) compared with the number of jobs or dollars that were supposed to be shipped in that time period. †

Cybermarketing

Any type of Internet-based promotion. Many marketing managers use the term to refer to any type of computer-based marketing. †

Cybernetic System

The information flow or information system (electronic, mechanical, logical) that controls an industrial process. †

Cybernetics

The study of control processes in mechanical, biological, electrical, and information systems. †

Data Collection

The act of compiling data for recording, analysis, or distribution. †

Data Dictionary

1) A catalog of requirements and specifications for an information system. 2) A file that stores facts about the files and databases for all systems that are currently being used or for the software involved. †

Data Mining

The process of studying data to search for previously unknown relationships. This knowledge is then applied to achieving specific business goals. †

Data Warehouse

A repository of data that has been specially prepared to support decision-making applications. Syn: decision-support data, information data warehouse. †

Database

A data processing file-management approach designed to establish the independence of computer programs from data files. Redundancy is minimized, and data elements can be added to, or deleted from, the file structure without necessitating changes to existing computer programs. †

Decision Support System (DSS)

A computer system designed to assist managers in selecting and evaluating courses of action by providing a logical, usually quantitative, analysis of the relevant factors. †

Die

A tiny square of silicon with an integrated circuit etched on it – more commonly known as a silicon chip. *

Discrete Manufacturing

The production of distinct items such as automobiles, appliances, or computers. †

Distributed Architecture

Software that runs simultaneously on different computers distributed throughout an organization, rather than on one central computer. *

Distributed Data Processing

A data processing organizational concept under which computer resources of a company are installed at more than one location with appropriate communication links. Processing is performed at the user's location generally on a smaller computer and under the user's control and scheduling, as opposed to processing for all users being done on a large, centralized computer system. †

Disintermediation

The process of eliminating an intermediate stage or echelon in a supply chain. Total supply chain operating expense is reduced, total supply chain inventory is reduced, total cycle time is reduced, and profits increase among the remaining echelons. See: echelon. †

Distribution

1) The activities associated with the movement of material, usually finished goods or service parts, from the manufacturer to the customer. These activities encompass the functions of transportation, warehousing, inventory control, material handling, order administration, site and location analysis, industrial packaging, data processing, and the communications network necessary for effective management. It includes all activities related to physical distribution, as well as the return of goods to the manufacturer. In many cases, this movement is made through one or more levels of field warehouses. Syn: physical distribution. 2) The systematic division of a whole into discrete parts having distinctive characteristics. †

Distribution Channel

The distribution route, from raw materials through consumption, along which products travel. Syn: marketing channel. See: channels of distribution. †

Distribution Requirements Planning (DRP)

1) The function of determining the need to replenish inventory at branch warehouses. A time-phased order point approach is used where the planned orders at the branch warehouse level are "exploded" via MRP logic to become gross requirements on the supplying source. In the case of multilevel distribution networks, this explosion process can continue down through the various levels of regional warehouses (master warehouse, factory warehouse, etc.) and become input to the master production schedule. Demand on the supplying sources is recognized as dependent, and standard MRP logic applies. 2) More generally, replenishment inventory calculations, which may be based on other planning approaches such as period order quantities or "replace exactly what was used," rather than being limited to the time-phased order point approach. †

Distribution Resource Planning (DRP II)

The extension of distribution requirements planning into the planning of the key resources contained in a distribution system: warehouse space, workforce, money, trucks, freight cars, etc. †

Domain Name

The unique name that identifies an Internet site. Domain names always have two or more parts separated by dots. The part on the left is the most specific and the part on the right is the most general. A given machine may have more than one domain name but a given domain name points to only one machine. †

Domain Name Service

A service used on the Internet to help the network route information to the correct computers. *

DRP

Abbreviation for distribution requirements planning. †

Dynamic Data

Data that can change constantly, such as the temperature of an item. *

Dynamic Lot Sizing

Any lot-sizing technique that creates an order quantity subject to continuous recomputation. See: least total cost, least unit cost, part period balancing, period order quantity, Wagner-Whitin algorithm. †

EAN International

The international group that administers bar code standards in many parts of the world. *

EDI

Abbreviation for electronic data interchange. †

EDI for Administration, Commerce, and Transport (EDIFACT)

A set of United Nations rules for electronic data interchange. These are international guidelines and standards for the electronic exchange of data regarding trade. †

Efficient Consumer Response (A)

1) A grocery industry-based, demand-driven replenishment system that links suppliers to develop a large flow-through distribution network. Information

technology is designed to enable suppliers to anticipate demand. Manufacture is initiated based on point-of-sale information. Accurate, instantaneous data are essential to this concept. 2) A management approach that streamlines the supply chain by improving its effectiveness in providing customer service and reducing costs through innovation and technology. †

Electronic Data Interchange (EDI)

The paperless (electronic) exchange of trading documents, such as purchase orders, shipment authorizations, advanced shipment notices, and invoices, using standardized document formats. †

Electrically Erasable Programmable Read-Only Memory (EEPROM)

A type of electronic memory that retains its contents even when the power is cut off and which can be reprogrammed. *

Electromagnetic Compatibility (EMC)

The ability of a system or product to function properly in an environment where other electromagnetic devices are used and not be a source itself of electromagnetic interference. *

Electromagnetic ID (EMID) Tag

A memory device with circuitry for communicating wirelessly with an external tag reader. An RFID tag is one type of electromagnetic ID tag. *

Electromagnetic Interference (EMI)

The effect one wireless systems or product has on neighboring systems or products. *

Electromagnetic Spectrum

The entire frequency range of electromagnetic waves. *

Electromagnetic Waves

Energy that is emitted in the form of waves. Types of electromagnetic waves include radio waves, gamma rays and x-rays. *

Electronic Article Surveillance (EAS)

Simple electronic tags that are either "on" or "off." When an item is purchased or borrowed legally, the tag is turned off. When someone passes a gate area holding an item with a tag that hasn't been turned off, an alarm sounds. *

Electronic Data Interchange (EDI)

A widely accepted method of sharing data over a business network. *

Electronic Product Code: (EPC)

The Auto-ID Center's coding scheme that will identify an item's manufacturer, product category and unique serial number. *

Encryption (A): Changing readable words into another form, called a cipher, that hides the text's meaning. †

Enterprise Resources Planning (ERP) System

1) An accounting-oriented information system for identifying and planning the enterprisewide resources needed to take, make, ship, and account for customer orders. An ERP system differs from the typical MRP II system in technical requirements such as graphical user interface, relational database, use of fourth-generation language, and computer-assisted software engineering tools in development, client/server architecture, and open-system portability. 2) More generally, a method for the effective planning and control of all resources needed to take, make, ship, and account for customer orders in a manufacturing, distribution, or service company. †

European Article Numbering (EAN)

The bar code standard used throughout Europe, Asia and South America. It is administered by EAN International. *

eXtensible Markup Language (XML)

A widely accepted way of sharing information over the Internet in a way that computers can use, regardless of their operating system. *

Expert System

A type of artificial intelligence computer system that mimics human experts by using rules and heuristics rather than deterministic algorithms. †

Extranet

A network connection to a partner's network using secure information processing and Internet protocols to do business. †

Family

A group of end items whose similarity of design and manufacture facilitates their being planned in aggregate, whose sales performance is monitored together, and, occasionally, whose cost is aggregated at this level. †

Feedback

The flow of information back into the control system so that actual performance can be compared with planned performance. †

Feedback Loop

The part of a closed-loop system that allows the comparison of response with command. †

File Transfer Protocol (FTP)

A protocol used to transfer files over the Internet. †

Finite Scheduling

A scheduling methodology where work is loaded into work centers such that no work center capacity requirement exceeds the capacity available for that work center. See: drum-buffer-rope, finite forward scheduling. †

Fluidic Self-Assembly

A manufacturing process, patented by Alien Technology, that involves flowing tiny microchips in a special fluid over a base with holes shaped to catch the chips. *

Frequency

The number of repetitions of a complete waveform in a specific period of time. 1 KHz equals 1,000 complete waveforms in one second. 1 MHz equals 1 million waveforms per second. *

Frequency Shift Keying (FSK)

A method of switching between different frequencies to transmit digital data. Often, one frequency represents a one, the other a zero. *

Global Trade Item Number (GTIN)

A superset of bar code standards, which is used internationally. In addition to manufacturer and product category, GTIN also includes shipping, weight and other information. The EPC is designed to provide continuity with GTIN. *

Graphical User Interface (GUI)

A connection between the computer and the user employing a mouse and icons so that the user makes selections by pointing at icons and clicking the mouse. †

Hardware

The physical, touchable, material parts of a computer or other system. Hardware for RFID systems consists of tags and readers, along with the computers required to collate, process and communicate the data generated. *

Heuristics

A form of problem solving in which the results or rules have been determined by experience or intuition instead of by optimization. Heuristics can be used in such areas as forecasting; lot sizing; or determining production, staff, or inventory levels. †

High-Frequency Tags

Tags operating in the 13.56 MHz range. *

Holonic Manufacturing System (HMS)

A method for manufacturing goods based on the cooperation of autonomous, functionally complete entities with diverse and often conflicting goals. Holonic manufacturing is still in the early stages of development, but can be greatly enhanced by RFID technology. *

HTML1

Abbreviation for hypertext markup language. †

Hybrid EDI

A situation in which only one trading partner is EDI enabled, while the other continues to use paper and fax. Usually the EDI-enabled partner would have electronic documents converted to fax. †

Hypertext

A system of relating information without using menus or hierarchies. †

Hypertext Markup Language (HTML)

A language used to create Web pages that permits the user to create text, hypertext links, and multimedia elements within the page. HTML is not a programming language, but a way to format text. †

Hypertext Transfer Protocol (HTTP)

A protocol that tells computers how to communicate with each other. Most internet addresses begin with http://. †

Independent Demand

The demand for an item that is unrelated to the demand for other items. Demand for finished goods, parts required for destructive testing, and service parts requirements are examples of independent demand. See: dependent demand. †

Industrial, Scientific, and Medical (ISM) Bands

A group of unlicensed frequencies of the electromagnetic spectrum. It isn't necessary to buy a license from the government before using communications equipment that operates at an ISM band frequency. *

Information System Architecture

A model of how the organization operates regarding information. The model considers four factors: (1) organizational functions, (2) communication of coordination requirements, (3) data modeling needs, and (4) management and control structures. The architecture of the information system should be aligned with and match the architecture of the organization. †

Information Technology

The technology of computers, telecommunications, and other devices that integrate data, equipment, personnel, and problem-solving methods in planning and controlling business activities. Information technology provides the means for collecting, storing, encoding, processing, analyzing, transmitting, receiving, and printing text, audio, or video information. †

Integrated Circuit (IC)

This is another name for a chip or microchip. ICs make up the brains of computers. *

Intelligent Agent

A program that regularly gathers information without the owner being present. †

Internet

A worldwide network of computers belonging to businesses, governments, and universities that enables users to share information in the form of files and to send electronic messages and have access to a tremendous store of information. †

Internet Protocol (IP)

The network layer for the TCP/IP protocol suite widely used on Ethernet networks. It routes packets of data among computers connected to a network. *

Interrogator

An RFID reader. *

Line-of-Sight Technology

Technology that requires an item to be "seen" to be automatically identified by a machine. Bar codes and optical character recognition are two line-of-site sight technologies. *

Logic Gate

Tiny switches on microchip circuits that enable the chip to perform certain operations. *

Low-Frequency Tags

RFID tags that communicate with readers at 125 KHz. *

Knowledge-Based System

A computer program that employs knowledge of the structure of relations and reasoning rules to solve problems by generating new knowledge from the relationships about the subject. †

LAN

Acronym for local area network. †

Learning Curve

A curve reflecting the rate of improvement in time per piece as more units of an item are made. A planning technique, the learning curve is particularly useful in project-oriented industries in which new products are frequently phased in. The basis for the learning curve calculation is that workers will be able to produce the product more quickly after they get used to making it. Syn: experience curve, manufacturing progress curve. †

Less than Truckload (LTL)

Either a small shipment that does not fill the truck or a shipment of not enough weight to qualify for a truckload quantity (usually set at about 10,000 lbs.) rate discount, offered to a general commodity trucker. †

Logistics

In an industrial context, the art and science of obtaining, producing, and distributing material and product in the proper place and in proper quantities. 2) In a military sense (where it has greater usage), its meaning can also include the movement of personnel. †

Logistics System

The planning and coordination of the physical movement aspects of a firm's operations such that a flow of raw materials, parts, and finished goods is achieved in a manner that minimizes total costs for the levels of service desired. †

Maintenance, Repair, and Operating Supplies (MRO)

Items used in support of general operations and maintenance such as maintenance supplies, spare parts, and consumables used in the manufacturing process and supporting operations. †

Manufacturing Execution Systems

Programs and systems that participate in shop floor control, including programmed logic controllers and process control computers for direct and supervisory control of manufacturing equipment; process information systems that gather historical performance information, then generate reports; graphical displays; and alarms that inform operations personnel what is going on in the plant currently and a very short history into the past. Quality control information is also gathered and a laboratory information management system may be part of this configuration to tie process conditions to the quality data that are generated. Thereby, cause-and-effect relationships can be determined. The quality data at times affect the control parameters that are used to meet product specifications either dynamically or off line. †

Manufacturing Planning and Control System (MPC)

A closed-loop information system that includes the planning functions of production planning (sales and operations planning), master production scheduling, material requirements planning, and capacity requirements planning. Once the plan has been accepted as realistic, execution begins. The execution functions include input-output control, detailed scheduling, dispatching, anticipated delay reports (department and supplier), and supplier scheduling. A closed-loop MRP system is one example of a manufacturing planning and control system. †

Manufacturing Resource Planning (MRP II)

A method for the effective planning of all resources of a manufacturing company. Ideally, it addresses operational planning in units, financial planning in dollars, and has a simulation capability to answer what-if questions. It is made up of a variety of processes, each linked together: business planning, production planning (sales and operations planning), master production scheduling, material requirements planning, capacity requirements planning, and the execution support systems for capacity and material. Output from these systems is integrated with financial reports such as the business plan, purchase commitment report, shipping budget, and inventory projections in dollars. Manufacturing resource planning is a direct outgrowth and extension of closed-loop MRP. †

Marketing Channel

That set of organizations through which a good or service passes in going from a raw state to the final consumer. Syn: distribution channel. See: channels of distribution. †

Mass Customization

The creation of a high-volume product with large variety so that a customer may specify his or her exact model out of a large volume of possible end items while manufacturing cost is low because of the large volume. An example is a personal computer order in which the customer may specify processor speed, memory size, hard disk size and speed, removable storage device characteristics, and many other options when PCs are assembled on one line and at low cost. †

Master Production Schedule (MPS)

The master production schedule is a line on the master schedule grid that reflects the anticipated build schedule for those items assigned to the master scheduler. The master scheduler maintains this schedule, and in turn, it becomes a set of planning numbers that drives material requirements planning. It represents what the company plans to produce expressed in specific configurations, quantities, and dates. The master production schedule is not a sales item forecast that represents a statement of demand. The master production schedule must take into account the forecast, the production plan, and other important considerations such as backlog, availability of material, availability of capacity, and management policies and goals. Syn: master schedule. †

Material Requirements Planning (MRP)

A set of techniques that uses bill of material data, inventory data, and the master production schedule to calculate requirements for materials. It makes recommendations to release replenishment orders for material. Further, because it is time-phased, it makes recommendations to reschedule open orders when due dates and need dates are not in phase. Time-phased MRP begins with the items listed on the MPS and determines (1) the quantity of all components and materials required to fabricate those items and (2) the date that the components and material are required. Time-phased MRP is accomplished by exploding the bill of material, adjusting for inventory quantities on hand or on order, and offsetting the net requirements by the appropriate lead times. †

Microchip

A microelectronic semiconductor device comprising many interconnected transistors and other components. Also called a chip or an "integrated circuit." *

Micron

A unit of length equal to one millionth of a meter or one thousandth of a millimeter. *

Model

A representation of a process or system that attempts to relate the most important variables in the system in such a way that analysis of the model leads to insights into the system. Frequently, the model is used to anticipate the result of a particular strategy in the real system. †

Modulation

Changing the frequency, phase, or amplitude of a wave to transmit data. *

Nanoblock

The term Alien Technology uses to describe its tiny microchips, which are about the width of three human hairs. *

National Stock Number (NSN)

The individual identification number assigned to an item to permit inventory management in the federal (U.S.) supply system. †

Network

Any system that transmits voice, video and/or data between users. *

Neural Network

A software system loosely based on how the brain works. It tries to simulate the multiple layers of elements called neurons. Each neuron is tied to several neighbors with a value that signifies the strength of the connections. Learning is accomplished by changing the values to cause the network to report appropriate results. Neural networks have been used for market forecasts and other applications. †

Object Name Service (ONS)

An Auto-ID Center designed system for looking up unique Electronic Product Codes and pointing computers to information about the item associated with the code. ONS is similar to the Domain Name System, which points computers to sites on the Internet. *

Object-Oriented Programming (OOP)

Within computer programming, the use of coding techniques and tools that reflect the concept of viewing the business environment as a set of elements (or objects) with associated properties, e.g., data, data manipulation/actions, inheritance. The objects encapsulate, through data and functions, the properties of the business that are of interest. †

Open Systems Interconnection (OSI)

A communication system where a user can communicate with another user without being constrained by a particular manufacturer's equipment. †

Operating System

A conglomeration of software that controls the hardware and application programs that perform the logical processing of the system. It is a system of programs that controls the execution of computer programs and may provide scheduling, accounting, debugging, and input/output control. †

Operations Research

1) The development and application of quantitative techniques to the solution of problems. More specifically, theory and methodology in mathematics, statistics, and computing are adapted and applied to the identification, formulation, solution, validation, implementation, and control of decision-making problems. 2) An academic field of study concerned with the development and application of quantitative analysis to the solution of problems faced by management in public and private organizations. Syn: management science. †

Part Coding and Classification

A method used in group technology to identify the physical similarity of parts. †

Part Family

A collection of parts grouped for some managerial purpose. †

Passive Tag

An RFID tag that does not use a battery. The tag draws energy from an electromagnetic field created by the reader. *

Phase Shift Keying (PSK)

A method of communicating information by switching transmission between different phases of the waveform to represent digital data. *

Physical Markup Language (PML)

An Auto-ID Center designed method of describing products in a way computers can understand. PML is based on the widely accepted eXtensible Markup Language used to share data over the Internet in a format all computers can use. *

PML Server

A dedicated computer that will respond to requests for Physical Markup Language (PML) files related to individual Electronic Product Codes. The manufacturer of the item may maintain the PML files and server. *

Process Industries

The group of manufacturers that produce products by mixing, separating, forming, and/or performing chemical reactions. Paint manufacturers, refineries, and breweries are examples of process industries. †

Process Manufacturing

Production that adds value by mixing, separating, forming, and/or performing chemical reactions. It may be done in either batch or continuous mode. See: project manufacturing. †

Product Family

A group of products with similar characteristics, often used in production planning (or sales and operations planning). †

Protocol

In information systems, a set of rules for defining the format and relationships for sharing information between devices. These rules govern the transmission of data across a network and serve as the grammar of data communication languages. †

Radio Frequency Identification (RFID)

A method of identifying unique items using radio waves. The big advantage over bar code technology is lasers must see a bar code to read it. Radio waves do not require line of site and can pass through materials such as cardboard and plastic. *

Radio Waves

Electromagnetic waves that fall within the lower end of the electromagnetic spectrum. *

Read Range

The distance from which a reader can communicate with a tag. Range is influenced by the power of the reader, frequency used for communication, and the design of the antenna. *

Read-Only Memory (ROM)

A form of storing information on a chip that cannot be overwritten. Read-only chips are less expensive than read-write chips. *

Read-Write

The ability to read and overwrite stored information. Chips for read-write RFID tags are more expensive than equivalent read-only chips. *

Reader

Also called an interrogator. The reader communicates with the RFID tag and passes the information in digital form to a computer system. *

Reader Collision

A problem that occurs when signals from readers with overlapping fields interfere with one another. *

RFID Transponder

See transponder. *

Savant

Distributed network software that manages and moves data related to Electronic Product Codes. *

Search Engines

Web software that enables a user to find a page or Web site devoted to a particular topic. †

Semi-passive Tags

RFID tags that use a battery to run the chip's circuitry, but communicate by drawing power from the reader. *

Sensors

Devices that can monitor and adjust differences in conditions to control equipment on a dynamic basis. †

Service Parts

Those modules, components, and elements that are planned to be used without modification to replace an original part. Syn: repair parts, spare parts. †

Server

A computer that processes and fulfills requests for files, Web pages or other digital information. *

Servo System

A control mechanism linking a system's input and output, designed to feed back data on system output to regulate the operation of the system. †

Setup Costs

Costs such as scrap costs, calibration costs, downtime costs, and lost sales associated with preparing the resource for the next product. Syn: changeover costs, turnaround costs. †

Shrinkage

Reductions of actual quantities of items in stock, in process, or in transit. The loss may be caused by scrap, theft, deterioration, evaporation, etc. †

Smart Cards

A broad term used for a plastic card (usually the size of a credit card) with an embedded microchip. Some smart cards contain an RFID chip so they can identify the holder without requiring any physical contact with a reader. RFID smart cards are often called "contactless" smart cards. *

Software

Also called a "computer program" or "program." Software is essentially the instructions that tell the physical computer – the hardware – what to do. Software can be written in different computer languages and generally falls into two categories: system software and application software or application programs. System software is any software required to support the production or execution of application programs but which is not specific to any particular application. Examples of system software would include the operating system and network software that directs traffic or checks passwords. Application software is the programs that run on top of the system software and perform specific functions, such as record keeping. *

Static Data

Data that doesn't change, such as facts relating to the material composition of a product. *

Structured Query Language (SQL)

A computer language that is a relational model database language. Such a language has an English vocabulary, is nonprocedural, and provides the ability to define tables, screen layouts, and indices. †

Supply Chain

The global network used to deliver products and services from raw materials to end customers through an engineered flow of information, physical distribution, and cash. †

Supply Chain Inventory Visibility

Software applications that permit monitoring events across a supply chain. These systems track and trace inventory globally on a line-item level and notify the user of significant deviations from plans. Companies are provided with realistic estimates of when material will arrive. †

Synthetic Polymers

Man-made compounds that make up plastic-like materials. Special types of synthetic polymers may one day offer an inexpensive replacement to silicon in microchips. *

Tag

The generic term for a radio frequency identification device. Tags are sometimes referred to as smart labels. *

Tag Collision

Interference caused when more than one RFID tag sends back a signal to the read at the same time. *

Task Management System

A method of organizing and customizing software to execute a set of tasks automatically. *

TCP/IP

Abbreviation for transmission control protocol/Internet protocol. †

Technical/Office Protocol (TOP)

An application-specific protocol based on open systems interconnection (OSI) standards. It is designed to allow communication between computers from different suppliers in the technical development and office environments. †

Temporal Data

Data that changes discretely and intermittently throughout an object's life, such as its location. *

The Internet Engineering Task Force (IETF)

An open, international group of network designers, operators, vendors, and researchers concerned with the evolution of the Internet architecture. *

Transmission Control Protocol (TCP)

A set of formal communications rules developed to internetwork dissimilar types of computers. TCP is the connection-oriented protocol built on top of Internet Protocol (IP) and is nearly always seen in the combination TCP/IP. It adds reliable communication and flow-control. TCP/IP has become the de facto standard for communicating over the Internet. *

Transmission Control Protocol/Internet Protocol (TCP/IP)

The communication protocol used by the Internet. †

Transponder

A radio transmitter-receiver that is activated when it receives a predetermined signal. RFID tags are sometimes referred to as transponders. *

Ultra-high Frequency (UHF)

The term generally given to waves in the 300 MHz to 3 GHz. UHF offers high bandwidth and good range, but UHF waves don't penetrate materials well and require more power to be transmitted over a given range than lower frequency waves. *

Unified Modeling Language (UML)

An open, standard method of modeling large, complex computer systems. *

Uniform Code Council (UCC)

The nonprofit organization that overseas the Uniform Product Code, the barcode standard used in North America. *

Universal Product Code (UPC)

The barcode standard used in North America. It is administered by the Uniform Code Council. *

URL

Abbreviation for uniform resource locator. †

User Datagram Protocol (UDP)

A set of communications rules that govern the transmission of data over a network. UDP doesn't require a connection or guarantee the delivery of data, so all error processing and retransmission must be taken care of by the application program. *

Value Chain

The functions within a company that add value to the goods or services that the organization sells to customers and for which it receives payment. †
value chain analysis (A): An examination of all links a company uses to produce and deliver its products and services starting from the origination point and continuing through deli1very to the final customer. †

Wafer

A small thin circular slice of a semiconducting material, such as pure silicon, on which an integrated circuit can be formed. Silicon wafers are usually eight to 12 inches in diameter. *

XML

See eXtensible Markup Language. *

XML Query Language (XQL)

A method of querying a database based on XML. Files created using the Auto-ID Center's Physical Markup Language can be searched using XQL. *

XML

Abbreviation for extensible markup language. †